AN INTRODUCTION TO ENGINEERING:
Methods, Concepts, and Issues

Board of Advisors, Engineering

An Introduction to Engineering: Methods, Concepts, and Issues

Edward V. Krick

Lafayette College
Easton, Pennsylvania

John Wiley & Sons, Inc.
New York London Sydney Toronto

Library of Congress Cataloging in Publication Data:

Krick, Edward V
An introduction to engineering.

Bibliography: p.
Includes index.
1. Engineering. 2. Engineering design. I. Title.
TA147.K72 620'.0023 75-41432
ISBN 0-471-50750-4

Printed in the United States of America

10 9 8 7 6 5 4 3 2 1

TO

Caroline, Jessica, and Kyle

TO THE TEACHER

If you are using or considering this book for a course, request a copy of the teacher's manual, which is more than a solutions manual. It includes a detailed syllabus for an introductory course, a film list, and additional problems. It will be particularly useful if you design problems. I present such problems in a manner that approximates real life, instead of assigning "Problem 2 at the end of Chapter 4." In the hope that you concur, in the manual I have included design problems in the form of letters, memoranda, and the like, which you can reproduce and utilize.

In using this book I urge you to assign *To the Student,* which outlines my objectives. Certainly the student should know the author's intentions. One objective is to introduce engineering to a person who is considering or beginning an engineering education. There are five ways of accomplishing this.

1. **Present case studies of engineering in action.**
2. **Describe the origin and nature of contemporary engineering.**
3. **Outline the important attributes of an engineer.**
4. **Describe the process of design, that is the sequence of activities that starts with recognition of a problem and ends with a functional, economical, and otherwise satisfactory solution. It includes problem definition, analysis, synthesis, invention, performance prediction, decision making, optimization, specification, and most other skills and techniques that are part of the engineering method. If you describe the design process, you describe the essence of engineering.**
5. **Involve the student in design.**

This book accomplishes the first four of these; an accompanying course can do the fifth.

Another objective is to improve the student's understanding of the purposes of an engineering education in terms of the skills, knowledge, and attitudes it will develop. It is not easy for lower classmen to visualize the ultimate purposes of what they are studying, making the first two years of an engineering program a motivational trial. This need not be so, and if freshmen are given the rationale for their curriculum, it won't be so.

If you are familiar with my earlier book, (*An Introduction to Engineering and Engineering Design,* Wiley, Inc., 1969) you are probably curious about what is different in this one. In *intent,* nothing is different; both are written to provide a solid foundation for the reader's development into an engineering professional. In *content,* however, there is a marked difference. In this book there are 57 pages devoted to the interaction between engineering and society, the professional implications of that interaction, and the issues raised, which is a significant increase over the nine pages given to the subject in the 1969 book. It is a striking increase that parallels the growing interest of engineers in such matters.

Some educators insist that engineering lower classmen, even upper classmen, are not ready for implications and issues. But I say that embryo engineers are concerned about such matters, and that it is a mistake not to give early indication that we care too. Furthermore, in a curriculum so steeped in analysis and the assimilation of factual knowledge, engineering students find this attention devoted to "issue-flavored" subject matter a refreshing change of pace.

I have labored to provide an unbiased characterization of the roles and influence of engineers in the modern world. However, this characterization may be positively biased in spite of my strivings. From someone as fond of engineering as I am, complete objectivity is probably too much to expect. There is a fine but crucial line between a harmless display of enthusiasm for one's profession and harmful distortion. I trust I haven't overstepped that line.

Never over the ten-plus years that I have been teaching an introductory course have I regretted moving into this phase of an engineering education. In fact, I find freshmen refreshing. They are imaginative (more so, I am afraid, than upper classmen), receptive, and appreciative—generally delightful to work with. I wish you the same rewards.

Edward V. Krick

TO THE STUDENT

My objectives in writing this book are to:

1. ***Contribute to a vital part of your general education.*** **Today, much in our lives is shaped and sustained by technology. No education is complete if it ignores the origins, workings, and widespread effects of this force. Although most of my objectives involve engineering, this one transcends career matters. It concerns your *general* preparation for a technically complex world. Take feedback systems, the subject of Chapter 12. An understanding of such systems was at one time appropriate only for the technically educated. Now, however, this should be common knowledge; feedback phenomena have become that pervasive in social, economic, political, and technical systems.**
2. ***Help you decide whether engineering is the right career for you.*** **To do this, I offer a clear picture of engineering—what it is, what it requires, what it offers, and what it lacks. However, this is a challenge, since there are many misconceptions to overcome. For instance (especially because of the news media), most people seem confused about the comparative roles of scientists and engineers. Also, many people can only picture an engineer "with hard hat and transit." (Now I grant you some engineers sometimes wear hard hats and occasionally look into a transit, but this is hardly a representative image.) These misconceptions are inevitable because the public rarely sees an engineer at work except when he is out in the field. Everyone sees the teacher and the doctor in action, but rarely the engineer. Yet, if you are to make an intelligent career decision, you must be aware of prevailing misconceptions. Alerting you to them is up to me.**
3. ***Start the development of certain mental skills that are important to your success in the engineering profession.*** **Much of this book deals with design, optimization, and modeling skills. Of the many skills employed in engineering practice, I emphasize these three because they are essential and because familiarity with them tells you a great deal about engineering.**
4. ***Acquaint you with basic technical terminology.*** **If you are to understand engineering literature and converse with its practitioners, you must be familiar with its special vocabulary.**
5. ***Cultivate a professional attitude, with emphasis on your sensitivity to the implications of engineers' creations.*** **This is an important and complex matter. It mainly involves attitudes and, since they are not ordinarily developed overnight, the process should begin early in your engineering education. And so a major thrust of this book is the exploration of the impact of engineering on the affairs and condition of man *and* its professional implications.**
6. ***Improve your understanding of the goals of an engineering education and the courses that it includes.*** **Engineering educators know what they are trying to develop in their students, so why shouldn't *you* know what those characteristics are? Among other things, knowing the objectives will enable you to become a more active participant in the process and thereby to get more out of it.**

I hope this book achieves these objectives for you.

Edward V. Krick

CONTENTS

AN INTRODUCTION TO ENGINEERING:
Methods, Concepts, and Issues

An Overview

PART 1

Through the medium of the printed page, you are going to witness engineers in action. This exposure will give you a general impression of what engineers do, the problems, challenges, and difficulties they encounter, and the qualities they must possess to perform their function.

Chapter 1

ENGINEERS IN ACTION—CASE HISTORIES

An engineer at Computer Electronics Company (C.E.C.) has conceived an information-processing system that could be very valuable to physicians and their patients. This system, called the Diagnosticator, is designed to assist in diagnosing illnesses. The physician would examine a patient and then feed his findings into the Diagnosticator. The machine would process this information and return a list of ailments and, for each ailment, it would give the probability of that ailment being the patient's trouble. For example, in response to a specific set of symptoms and certain information about the particular patient the machine says that there are 63 chances out of 100 that the patient has ailment A, 18 out of 100 that his illness is B, and so on.

Although the management of C.E.C. appreciates the potential usefulness of the Diagnosticator, it will not invest a substantial sum in this venture unless it is convinced that sale of such machines will yield sufficient profits and other benefits to justify the total cost of developing, manufacturing, and marketing them. On the basis of the promise of this idea, management has commissioned the engineer to provide preliminary specifications of this device and a forecast of the total cost of getting it on the market. Based on this "first approximation," if it appears that the project will ultimately yield an attractive return, the engineer will be authorized to proceed with the detailed design and operational testing of the Diagnosticator.

Among other things, management has decided that the device must produce the desired diagnostic information within one minute, must not be cumbersome, and must operate on ordinary house current. Preliminary specifications and predictions must be completed within two months.

COMMENTARY. The engineer must find the most effective means of transforming patient information into a set of ailment probabilities. The device must satisfy certain restrictions (e.g., it must produce results within one minute) and must be superior with respect to certain criteria (e.g., manufacturing cost, appeal to potential buyers). Two months are allotted for the project. This, in brief, is his problem.

During the project the engineer uses his knowledge and inventiveness to devise a variety of possible solutions. One solution would be to locate a number of typewriter-size electronic devices in users' offices (Figure 1-1). With these devices, medical personnel would transmit symptom information to, and receive results from, a centrally located information-processing unit that serves many customers. An alternative plan is to provide each user with an independent unit. Some of the relative merits of these alternatives are obvious, but what these differences mean in dollars is not as apparent.

The engineer is also investigating various methods of entering patient information into these machines and of getting results out, alternative ways of processing the data, and different types of components. Many alternative solutions are feasible, but they are not equally desirable. The engineer must evaluate competing schemes and come up with the one "best" system.

Throughout this project he works with many people —mathematicians (about probability theory), marketing specialists, manufacturing experts, and medical people. He spends a great deal of time observing physicians in action and querying them about their needs, preferences, and reactions to various ideas. He has more contact with people than he had anticipated before his first engineering job.

After his investigation, the engineer will present his recommendations to top executives in an extensive written report which is accompanied by an oral briefing. The report will include sketches of the proposed system (Figure 1-1), descriptions of its operation and performance, anticipated costs, and other economic projections that management needs to make an intelligent decision. The future of the project depends on the information received by management.

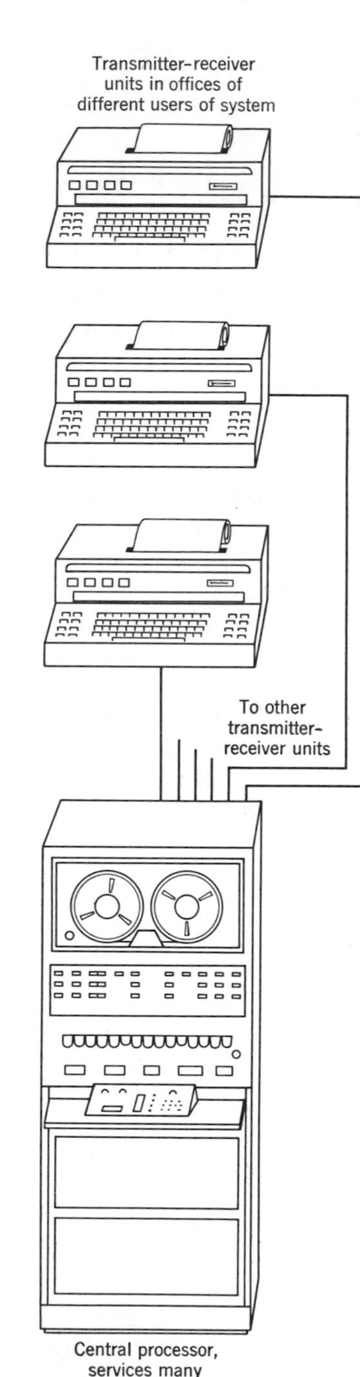

Figure 1-1 *The engineer prepared sketches of the major alternatives he is considering for inclusion in his report to management. This is his sketch of the centralized Diagnosticator scheme that services terminals in users' offices.*

An Automatic Production Machine

A large telephone system includes millions of switches. Engineering has made it possible for those switches to function for years, making billions of connections without wearing out. But they are still subject to failure caused by moisture or foreign matter. The cost of isolating and remedying these failures plus the cost of preventive maintenance have concerned the company for years. An engineer was asked to reduce these costs by improving the reliability of the system.

During his investigations, he developed and evaluated many possibilities; finally, he selected the ingenious switch described by Figure 1-2 as the most promising solution. This device is remarkably fast, very reliable, maintenance-free, and superior to any previous switch. A very important question, however, that determines whether this novel switch can be useful to the company and its customers is: can it be economically manufactured by the millions?

To answer this question, a team of engineers was assigned the task of developing, if feasible, an economical method of making these switches. The solution is the remarkable machine shown in Figure 1-3.

<u>*COMMENTARY.*</u> This team's problem was to find the most economical means of transforming glass tubing, reeds, and gas into the specified switches—by the millions. It didn't take the team long to realize that making millions of switches by hand would be prohibitively expensive. The fate of this clever switch hinged on the team's ability to develop an economical machine—or a combination of operator and machine—to manufacture it. Of course, developing a method that places these reeds in glass tubing and aligns them to the close tolerances required is a challenge.

Throughout this project, its economic feasibility was under scrutiny. The team periodically paused to reevaluate the probability of developing an economical solution. If, at any time during the project, it appeared that all methods of manufacture would be too costly, the originator of this switch would have been looking

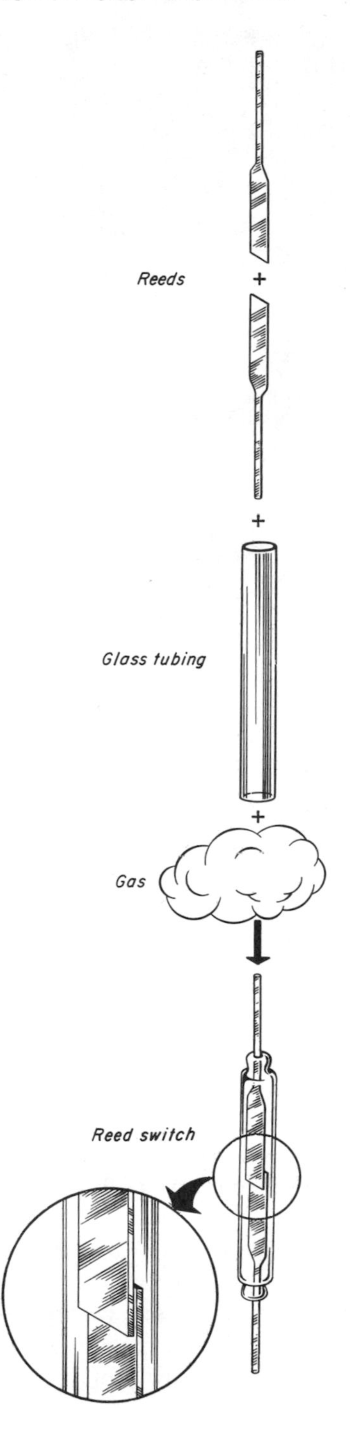

Figure 1-2 *The guts of this switch are two metal reeds sealed in glass tubing along with an inert gas. A close-up view shows the contact point where reed alignment and spacing are critical. In use, this assembly is mounted in an electromagnetic coil that, when energized, causes the reeds to make contact and complete a circuit.*

Figure 1-3 *An automatic machine that makes more than a million reed switches a year. It operates on the merry-go-round principle. The turret containing 18 identical assembly heads revolves and, as it does so, the reed switch gradually takes shape. At sequential stations around the turret's periphery, tubing is placed, reeds are inserted and aligned, the gap is set, gas is injected, the tubing is sealed, and the finished switch is ejected. The switches then go to the testing section of the machine where their physical and electrical characteristics are measured. Unsatisfactory switches are rejected and, on the basis of these measurements, the machine adjusts itself to correct whatever is causing the defects. (Picture credits begin on page 353.)*

for another solution to the problem—a switch cheaper to manufacture.

The switch designer specializes in developing devices used in the telephone system. The engineers subsequently assigned to the project develop the means to manufacture these devices. They are known as *process* or *manufacturing engineers*. Members of such design teams are usually experts in complementary fields. In this particular case, one engineer specializes in the behavior and forming of glass, another in machine components and mechanisms, another in electrical and magnetic phenomena, and

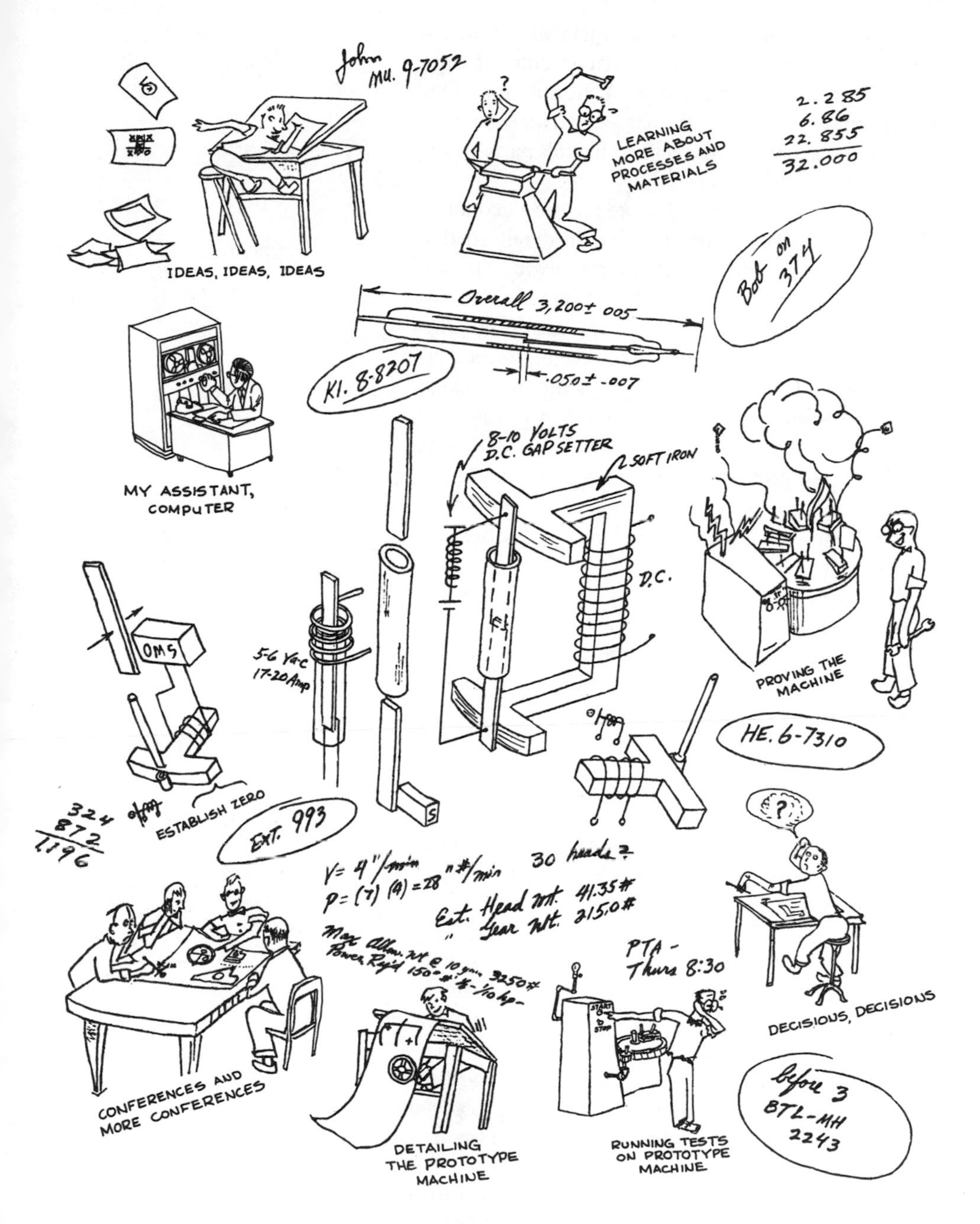

Figure 1-4 *A page from the notes of one of the team's engineers reflects, by way of his doodlings, the many alternatives, false starts, meetings, details, procedural steps, and communications involved in the development of the machine shown in Figure 1-3.*

so on. A close working relationship among members of an engineering team is vital, since each specialist concentrates on a facet that interacts with most other facets of the solution. One member of the group (referred to as the *project engineer*) serves as a coordinator of the group's activities, so that all parts of the final system are properly interrelated.

When the team believed that it had devised the most economical method, its proposal had to be specified in detail so that technicians and craftsmen could construct a prototype. The engineers were responsible for overseeing the construction of this prototype. They made modifications of their original design during the construction. When the model was complete, they supervised test runs of the machine; as a result, additional design modifications were instituted. Finally, after much testing and refining, the proposed machine was ready. Complete specifications of the final prototype were prepared by draftsmen so that additional machines could be constructed. Then a more effective switch became available for general use at a rate of many millions per year.

The task still was not complete. The process engineers followed up by observing the machine in use, recommending appropriate design changes, and evaluating the design so that future projects could benefit from experience with this machine.

A Domestic Water Desalter

Because of the diminishing supply of fresh water and the rapidly increasing demand for it, the problem of providing adequate amounts of drinking water is a pressing one. Developing economical sources of drinkable water is an *engineering* problem.

In many areas of the world, a promising source of drinking water is brackish water that lies underground. In anticipation of both the commercial and the service-to-mankind opportunities in this activity, a company is developing equipment that will convert such water to a usable form. One of these projects is the development of a converter that can be used in the home. The engineer responsible for this is now evaluating a promising prototype (Figure 1-5); it both purifies and desalts the water. The input is salt, brackish, or otherwise impure water; the output is demineralized and potable. Such a converter would be useful in homes, small commercial establishments, military field units, and on ships. It is effective and simple, needs little maintenance, and does not require a pressure vessel.

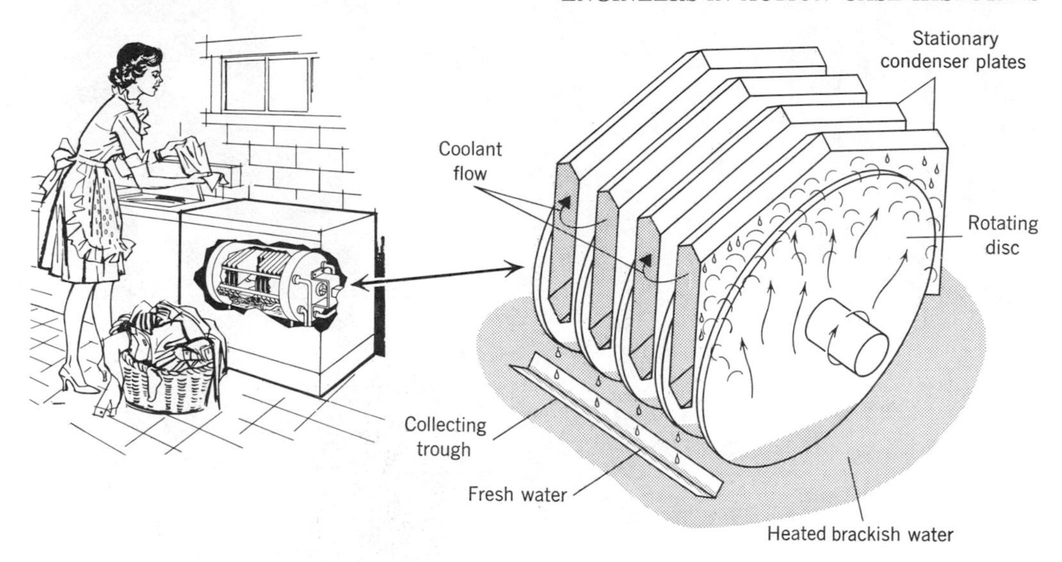

Figure 1-5 *Household unit for converting brackish or seawater. Close-up view shows the actual conversion mechanism. As the shaft rotates, the discs connected to it are coated with a thin film of warm, brackish water at the bottom of the tank. This thin film of water vaporizes as it passes through the air and condenses on the cool stationary plates. This condensate, now fresh water, drops into the collecting troughs.*

COMMENTARY. This is not simply a matter of finding *a* method of converting brackish water to potable form; after all, distillation has been known for centuries. The problem is to find a means of transforming *large quantities* of water at *a reasonable cost.*

The development of this converter is based partly on the engineer's technical and scientific knowledge and partly on her inventiveness. She could not have developed this machine without understanding vaporization and condensation phenomena, the behavior of thin films, thermal processes, and other scientific facts. This knowledge alone, however, could not have yielded this converter. The rotating discs interspersed with stationary collector plates, the particular configuration of these plates, and other unique features of this mechanism are *inventions*. These she did not find in handbooks or textbooks; they are fruits of her creative powers.

Real engineering effort and talent have gone into this machine. Many conversion schemes were evaluated, hours of tests were made, and considerable research was necessary. The result of this extensive development process is a well-engineered device that will probably prove financially successful and valuable as a service to the public.

(a)

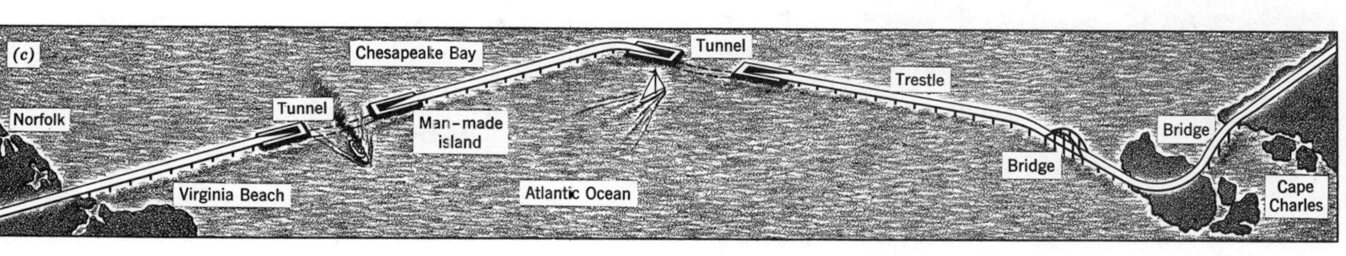

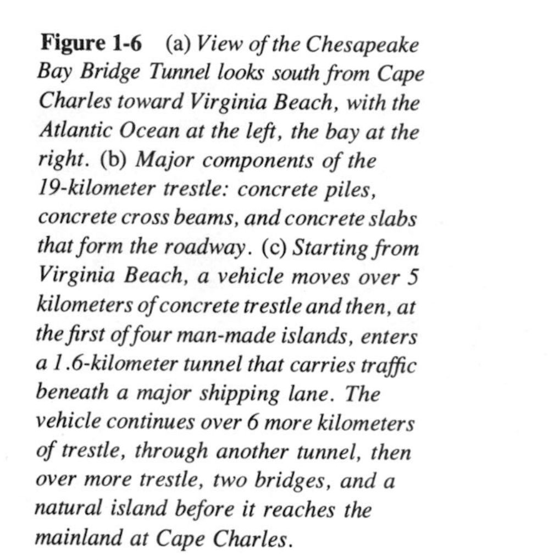

Figure 1-6 (a) *View of the Chesapeake Bay Bridge Tunnel looks south from Cape Charles toward Virginia Beach, with the Atlantic Ocean at the left, the bay at the right.* (b) *Major components of the 19-kilometer trestle: concrete piles, concrete cross beams, and concrete slabs that form the roadway.* (c) *Starting from Virginia Beach, a vehicle moves over 5 kilometers of concrete trestle and then, at the first of four man-made islands, enters a 1.6-kilometer tunnel that carries traffic beneath a major shipping lane. The vehicle continues over 6 more kilometers of trestle, through another tunnel, then over more trestle, two bridges, and a natural island before it reaches the mainland at Cape Charles.*

(a)
(b)

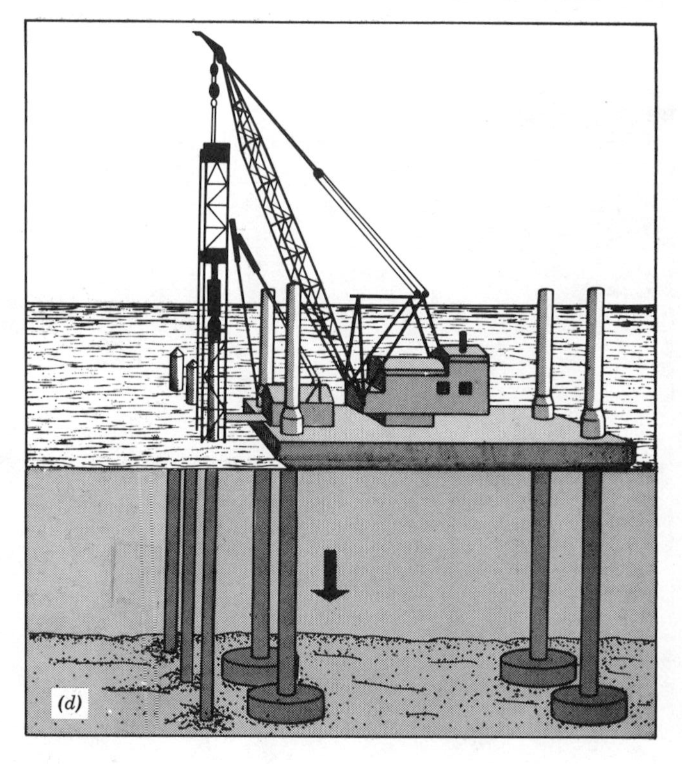

Figure 1-7 *The trestle was constructed by driving long (up to 52 meters) concrete piles deep into the bed of the bay by means of a specially designed machine nicknamed Big D, shown in* (a) *and* (b). *This $1.5 million rig is an enormous pile driver and crane mounted on a barge with retractable legs. Big D floats* (c) *into position and then becomes a stationary platform by extending its legs and hydraulically lifting itself out of the water* (d). *It ''stands'' in this position until its work is complete at that spot. Without this stability Big D would never be able to locate the huge piles to close tolerances.*

(a)

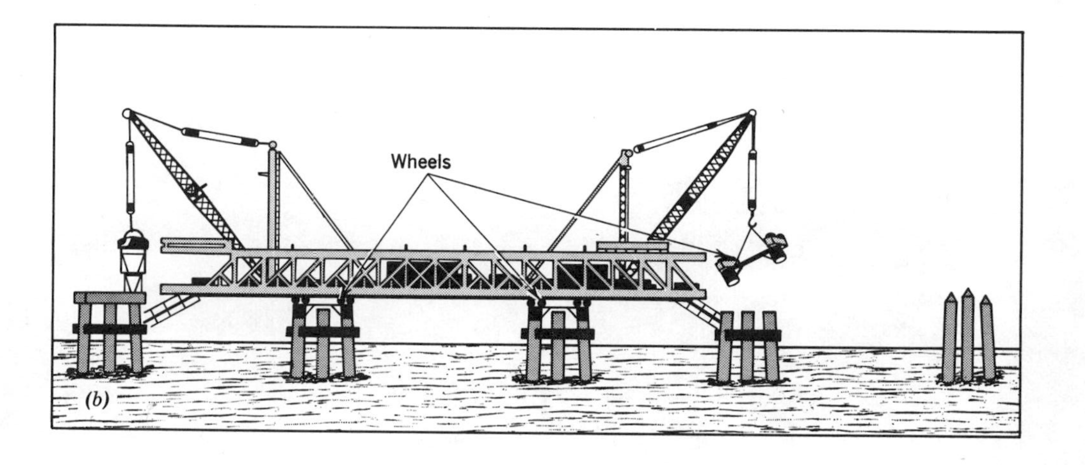

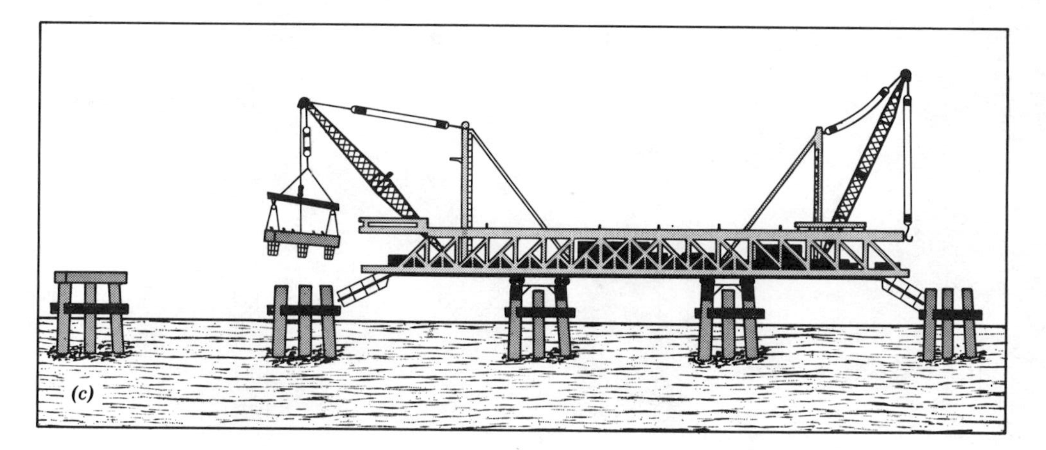

Figure 1-8 *Big D is followed by another special rig called the Two-Headed Monster, shown in* (a). *As it crawls along the tops of the piles, the leading "head" of the Monster trims the piles to a uniform height. Cross beams are placed by the Monster's trailing "head."*

The Monster moves along on wheels that are temporarily mounted on the tops of the piles. It has an extra set of wheels that it places on the next row of piles ahead of it (b), *after which it rolls forward to its next working position* (c).

(a)

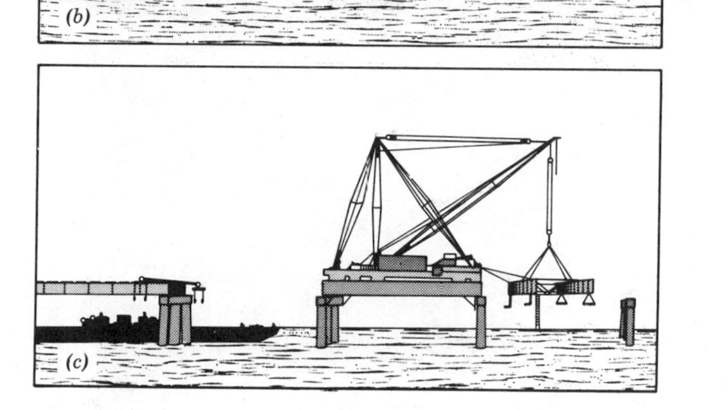

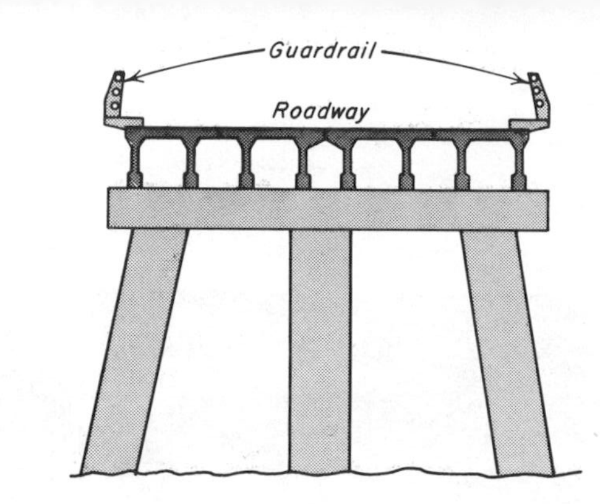

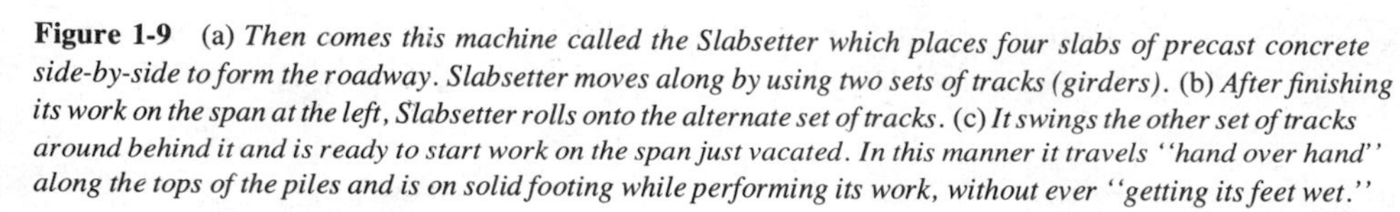

Figure 1-9 (a) *Then comes this machine called the Slabsetter which places four slabs of precast concrete side-by-side to form the roadway. Slabsetter moves along by using two sets of tracks (girders).* (b) *After finishing its work on the span at the left, Slabsetter rolls onto the alternate set of tracks.* (c) *It swings the other set of tracks around behind it and is ready to start work on the span just vacated. In this manner it travels "hand over hand" along the tops of the piles and is on solid footing while performing its work, without ever "getting its feet wet."*

Figure 1-10 *In the foreground is a man-made island that provides protection for the tunnel where it surfaces to connect with the trestle. The island at the other end of this tunnel is visible in the background. Each island is about 500 meters long and required almost 2 million tons of material. The island's sand core is protected by a heavy armor of stone.*

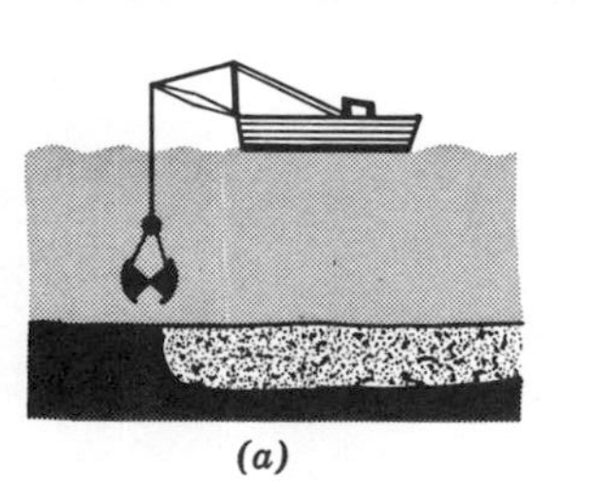
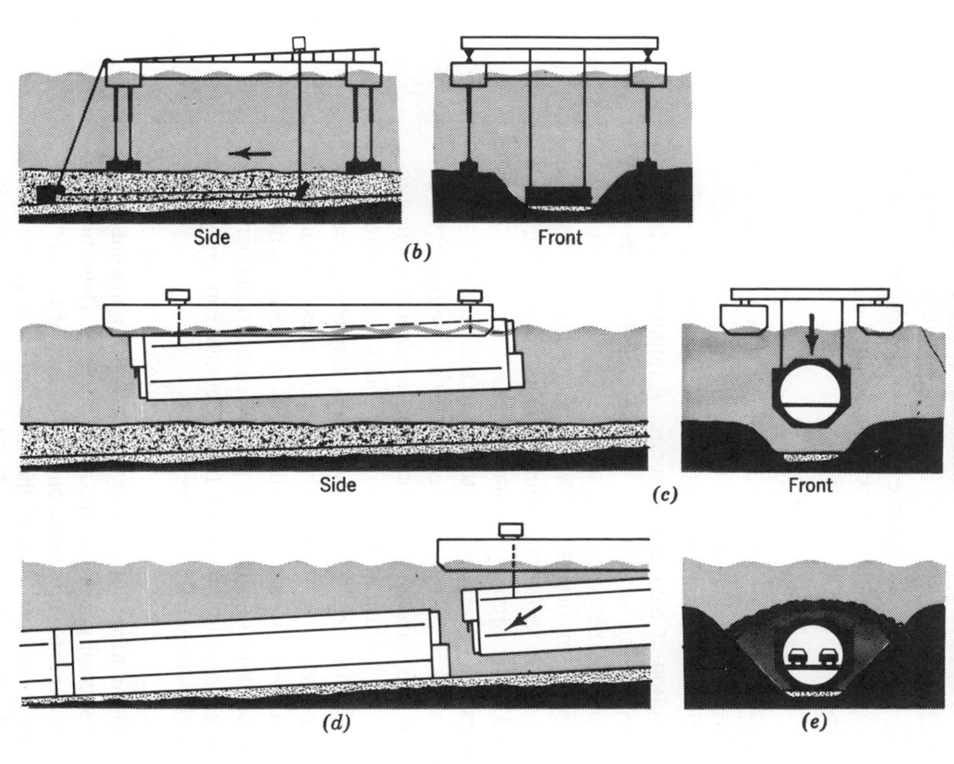

Figure 1-11 *Construction of the tunnels beneath the shipping channels is another interesting story.* (a) *First a dredge cut a trench roughly 30 meters wide and anywhere from 15 to 30 meters deep.* (b) *Then a device called a screed graded the bottom of this trench to within a few centimeters of the specified level. Note that the screed can be pitched to parallel the desired slope of the finished trench.* (c) *Next, a 90-meter prefabricated tube was lowered and positioned by a special barge, shown here.* (d) *It was then connected to the previously laid tube section.* (e) *Later the tubes were covered with layers of rock and sand, the joints sealed, the temporary end-seals removed, and the interiors finished. The result was one continuous submarine tube linking the man-made anchor islands.*

Why was this intensive engineering effort necessary for such a simple device? The simplicity is deceiving; it leads you to greatly underestimate the effort, ingenuity, analytical work, and investigation that went into this device. Without this work, the result would probably be more complicated and therefore more impressive to the untrained eye—but no more effective. In fact, a more complicated machine would be more susceptible to breakdown, more costly to make, and perhaps too expensive to sell.

The Chesapeake Bay Bridge-Tunnel

The numerous trestles, bridges, and tunnels of the $140 million structure in Figure 1-6 carry vehicles across the 29-kilometer mouth of Chesapeake Bay. The bridge-tunnel is a remarkable engineering achievement. It is the longest man-made fixed crossing of navigable ocean, built to withstand punishing waves and rushing tides.

COMMENTARY. As is often the case, this span was designed by an engineering consulting firm that specializes in such structures. The firm was commissioned to locate, design, and supervise construction of the entire structure. An unusual restriction was imposed on this structure: it could not pass over the main shipping channels, because a bridge above them could be bombed and trap navy vessels in the bay. Therefore, it was necessary to go under the channels, employing tunnels where bridges would have been placed under ordinary circumstances.

Engineers must give careful attention to the means by which their creations will be built. Especially in a case like this, the construction methods are as important as the design itself. Figures 1-7 to 1-11 illustrate this point. In particular, notice that trestle piles, cross members, roadway slabs, tunnel sections, and other components were prefabricated by mass production methods on land, where construction is achieved with less difficulty and expense. The use of prefabricated components is a method of construction resulting from adequate forethought on the part of the structure's designers.

The interrelationship between a structure's physical characteristics and the means of constructing it is also apparent if you look at the special construction equipment used. Determination of what equipment was needed and the design of it were major parts of the problem. In fact, the investment in construction equipment for this project was $15 million. Thus, the feasibility

Figure 1-12 *A VTOL hovering in a stationary position. This plane can move straight up and down or hover like a helicopter by using the three lift fans–one in each wing and one in the nose. Once airborne, the pitch of louvers under the wing fans is changed to deflect the fan exhaust rearward, thus producing horizontal acceleration. When the aircraft speed is sufficient for normal wing-supported flight, the fan louvers are closed and the craft functions as a conventional, highspeed jet plane, cruising at about 800 km/hr. The fans are driven by the same two jet engines that propel the plane in horizontal flight.*

of such a venture depends heavily on the engineers' ability to formulate a design that minimizes construction costs while satisfying all functional requirements.

Aircraft Development

For four years a design team has been developing a new airplane called a VTOL (pronounced *veetall,* meaning vertical takeoff and landing plane). The results of their efforts are described in Figures 1-12 to 1-16.

COMMENTARY. This team has developed a plane with sufficient thrust to raise it vertically and to propel it horizontally at competitive speeds. This is a challenge if you don't want the plane to be all engine. Another challenge arises from the tendency of such planes to tilt when in a hovering position, and to drift horizontally. To overcome these tendencies, a sophisticated control system is needed to maintain the plane's hovering stability. In VTOL design, problems associated with instability and its con-

Figure 1-13 *A multiple-exposure photograph of the VTOL taking off, showing the transition from vertical to horizontal flight.*

Figure 1-14 *A one-sixth scale model of the VTOL being readied for wind tunnel tests. The tufts of wool attached to the fuselage show airflow. The cables leading into the rear of the model bring power in and information out from measuring instruments.*

Figure 1-15 *This is an experimental model of the type used in outdoor tests of preliminary VTOL designs. In this case the designers are evaluating the hover and control capabilities of a plane with three small jet engines at each wing tip. Eventually they removed the ''leashes'' and allowed this rig to rise to a prudent altitude. Crude as this flying framework may seem, it told the engineers what they needed to know–without unnecessary fanciness, time, and expense.*

Figure 1-16 *The stakes are more than military contracts; naturally the company hopes to penetrate the commercial market with a VTOL. Therefore the design team has been considering transport VTOLs like this one, based on the concepts that they have developed for the military version shown in Figure 1-12.*

trol are some of the most challenging, requiring considerable mathematical skill and frequent use of computers.

The team experimented extensively with various models (Figures 1-14 and 1-15) in order to evaluate alternative designs. These engineers are skilled in instrumenting a model to obtain the necessary measurements. They also are capable of planning efficient experiments and intelligently interpreting the data that the experiments yield.

Even though working models have been testflown for many hours, the company still has no definite customer. The army financed the project, but the company has no guarantee of ever selling a VTOL. What the company receives in the future for its efforts depends on how well these seven engineers do their job and on competitors' designs.

Ventures like this are common. A company sees a future opportunity in an engineering creation that presently is beyond the state of the art. It develops its technical competence in the area by assigning engineers to design one or more experimental models. Such development efforts are sometimes funded by the government; at other times, the cost is borne by the company in the hope that the investment will lead to profitable contracts someday. If no contracts arise, the company may be able to recover its investment through commercial sales. Figure 1-16 indicates that the team is already thinking of this possibility.

A company usually carries on such projects at the same time that their competitors are working on similar projects. In each instance, several companies are building technical competence, gambling that it will eventually pay off. This adds excitement to the project, but it also affects job security.

Relate these cases to the remainder of the book; they provide illustrations for the generalizations, descriptions of method, discussions of impact, and other contents that follow. I am not exemplifying *the* practice of engineering through the preceding case studies. Since engineering is multifaceted, it is inevitable that there are roles and activities not represented here.

EXERCISES

1. Describe an engineering project; include the circumstances that gave rise to the project, the main challenges, the final result, and the ensuing effects. Some possibilities for which there is adequate resource material are: Aswan High Dam on the Nile;

relocation of the Temple of Abu Simbel; San Francisco Bay area rapid transit system (BART); computer-based airline reservation system; communications satellite; nuclear power plant; Verrazzano Narrows bridge; trans-Alaska pipeline; and trans-oceanic submarine cable.

2. Cultivate the ability to seek patterns in your experiences. If you think about it, you will probably observe that the better teachers and students excel at recognizing patterns. To develop your pattern recognition skills, search through the case histories of this chapter for characteristics common to most or all of the projects. These characteristics can be stated, implied, or assumed. For example, each project involves "hardware" (devices, machines, and structures).

WHAT DOES IT TAKE? **Chapter 2**

What is required to design the devices, machines, and structures that are described in Chapter 1? From the following generalizations about the problems engineers solve, *you* can infer many of the qualities possessed by competent practitioners. In this respect the case histories of Chapter 1 will be particularly useful.

An engineering problem usually begins with the recognition of a need or want that apparently can be satisfied by some physical device, structure, or process. At this stage things are likely to be vague. To illustrate, a shipbuilding company has tentatively decided to market an air-cushion vehicle (ACV) to compete with similar vehicles already available from other companies (Figure 2-1). In broad terms the management has decided on desired features of this vehicle, such as the price range, seating capacity, and cruising speed. The engineers now have their assignment: design a vehicle that satisfies the general performance characteristics given. This is typical of engineering assignments. The engineer is given the function or purpose to be served and probably some requirements and preferences for a solution. Such functional and performance specifications are usually selected by the engineer's superiors or clients often in collaboration with him or her.* The challenge to the engineer is to translate the loose statement of what is wanted into the specifications for a satisfactory means of fulfilling that objective.

In each of the Chapter 1 case histories there are numerous ways of achieving the specified objective; many, if not most of them, are not obvious, perhaps not even known, at the start of the

* I am in an awkward position. If I repeatedly fail to acknowledge that the engineer might be a woman, many readers will be justifiably upset. Yet this he-or-she, him-or-her gets tiring to you and to me. Consequently, unless this defect in our language is remedied before time of publication, when you find "he," the possibility of "she" is always implied. Forgive me, girls.

Figure 2-1 *An ACV is supported by a cushion of air trapped within its flexible skirts, enabling it to travel over land and water with ease. The vehicle shown carries 265 passengers and 30 automobiles, cruises at 100 kilometers per hour, and can negotiate 3-meter waves. Craft of this type provide ferry service across the English Channel.*

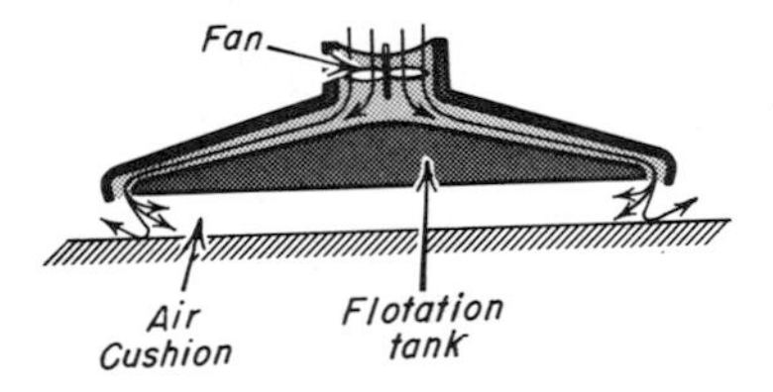

project. This is almost invariably true. It is up to the engineer to uncover and explore a number of the possibilities. The knowledge gained through education and experience is an important source of such solutions, but not the only one; the engineer must also rely heavily on ingenuity.

An air of urgency surrounds almost all engineering projects. Even if no target date has been set for a solution, there usually is pressure to produce results as quickly as possible. Consequently, the engineer generally must recommend a solution long before he has had time to uncover all the possibilities. Furthermore, instead

Figure 2-2 *In the design of this antenna, which is used to track and communicate with space vehicles, a number of conflicting objectives had to be resolved.*

of an exhaustive evaluation of the alternatives—something he simply does not have time for—he must rely heavily on judgment. This is a demanding aspect of an engineer's day-to-day work.

In most engineering problems there are conflicting objectives. The designer of the tracking antenna, pictured by Figure 2-2, knows that very well. Among the conflicting objectives he had to deal with were two major ones: long range and precise tracking ability. Very early in the project it became evident that an attempt to increase the antenna's range by enlarging the reflector would decrease its ability to stay on target because wind-caused vibration would be increased. True, the vibration could be overcome by stiffening the reflector with more steel, but that would add to its inertia which in turn would aggravate the aiming problem. Therefore, the designer had to make some difficult compromises in order to achieve a satisfactory balance between range and aimability.

And so it goes with every problem; as the engineer attempts to achieve "the ultimate" solution with respect to one performance characteristic, he is likely to lose ground with respect to other objectives. In the end a balance must be achieved, which is rarely simple.

The extent to which economic factors enter into engineering work cannot be overstated. If an engineer's creations are to benefit society, the intended users must be able to afford them. Therefore the engineer must have a keen interest in the costs of developing, producing, distributing, and using his solutions. Consider production cost—it depends mainly on the designer; he had better be concerned about it. The desalinator's ultimate cost to the consumer and, in fact, the economic success of the whole venture depend heavily on the producibility of the engineer's design. So it is with all engineering creations.

The engineer's concern with production and other costs is not solely for the benefit of the ultimate buyer. The survival of many businesses depend on the cost consciousness of their engineers. Moreover, whether or not an engineer's project (e.g., the Diagnosticator) ever gets off the ground depends on the economic promise of the basic concept. Private enterprise does not ordinarily embark on a venture unless it promises an attractive profit. In public ventures, too, a satisfactory benefit-to-cost ratio is demanded. Even though an engineer's proposal may satisfy its intended function completely, it is doomed if it will not yield a net gain to the business or community.

As you see and read about remarkable and often exciting engineering achievements, don't kid yourself! Not all of the engineering work behind them is challenging and sophisticated. Some of it is unglamorous, detailed, and tedious, as is true in any occupation. There are many details to be worked out and they had better be correct. An engineer doesn't endear himself to his employer by specifying the wrong screw size for a new-model TV set—causing the production department to stock 600,000 screws that don't fit! If errors in details can be this consequential, imagine what major blunders can cost! The accuracy of specifications is an engineering *must*.

Most of the creations described in this book are complex systems; they involve many components, and the interrelationships between them are complicated. Although it is not particularly large, the VTOL is a very complicated mechanism. This is a conspicuous trend in engineering creations—they are becoming more complicated, larger-scale, and more ambitious. Because of this complexity and the broad range of knowledge required in such projects, many engineering problems are handled by teams of engineers with varying backgrounds. The situation in which one engineer completely designs a device or structure is becoming rare. In fact, hundreds of engineers are involved in the design of a spacecraft. They are divided into teams with one team designing the propulsion subsystem, another the guidance subsystem, and so on for more than a dozen, major subsystems.

Generally, the result of an engineer's efforts is tangible—a physical device, structure, or process—as illustrated by the Diagnosticator, the desalinator, the reed switch machine, the bridge-tunnel, and the VTOL. This fact is probably the basis of a common misconception about engineering. Since the result is a device, structure, machine, or mechanism, many people conclude that engineers spend most of their time working with these things—like the mechanic, the television repairman, or the laboratory technician—but this is not ordinarily so. For instance, an engineer usually does much of his problem solving in the abstract. He works much more with information (e.g., fact gathering, computing, thinking, and communicating) than with things. Furthermore, technicians are usually employed to construct models for testing and demonstrating the engineer's creations, so he has little occasion to "work with his hands." In this respect, engineering work is quite different from what many people imagine it to be.

This is true for another reason: the typical engineer spends more time with people and people-problems than you might think. The frequent use of teams and the amount of specialization that prevails in practice are two reasons that contact with people requires more of an engineer's time (and sitting in solitude at the drawing board less) than you probably realize. There are those days when most of his time is spent making inquiries, issuing instructions, answering questions, providing advice, exchanging ideas, and seeking approvals. Consequently, the ability to mesh with other people in a variety of capacities is important; so is the ability to communicate effectively.

The engineer's involvement with people does not end here. A significant part of his work is the detection and appraisal of human needs and desires (e.g., the need for new sources of fresh water and the types, capacities, and number of water purifiers required). He must also be concerned with public acceptance of his solutions and, therefore, must get to know how people will use his creations, how they will react to them, and the features they prefer. He must also anticipate and be concerned about the effects of his creations on people. Thus the engineer is deeply involved with social needs, as well as with social acceptance and social effects of his creations.

So What Does It Take?

Skills

What is required to solve problems such as those in Chapter 1? First, you need problem-solving ability. This is only one of many skills that you must acquire. I have selected three of them for discussion in Part II: problem solving, modeling, and optimization.

These are not the only skills, however. Among other things, you should know how to prepare, conduct, and evaluate experiments that will give you a maximum amount of information with a minimum of time and expense. In experimentation and other phases of this work, you will rely heavily on measurement skills, and there is a lot to this matter of measurement. Closely related to measurement and experimentation is the ability to reach intelligent conclusions from observations; skillful interpretation of them is not as straightforward as you might think. This is true because of the uncontrollable variations in the characteristics of all materials, objects, and devices; also, *no* measurement system is perfect; furthermore, most of the engineer's conclusions must be based on relatively small numbers of observations. These circumstances complicate the process of drawing conclusions. Therefore, you must learn the potential sources of error in reaching conclusions, the limitations of small samples, the roles of chance, uncertainty, and prejudice, and the importance of carefully evaluating the reliability of available evidence. Engineers have an expression for it: "the art of skillful approximation."

Recall the engineer developing the Diagnosticator. During this project he was a frequent visitor to medical libraries, searching and reading books and papers on diagnostics. This required a certain bibliographic know-how, which he has had need for on numerous occasions and, certainly, he is no exception. Skill in the use of information resources (books, reports, articles, patents) is becoming more and more important for engineers. With knowledge expanding so rapidly, the desirability and the difficulty of searching for know-how relating to a specific problem increase. Without this searching skill, an engineer can miss a lot of valuable information and waste time in the process.

Thought skills are obviously important in engineering, and so a major goal of an engineering education is to sharpen your reasoning and your analytical and other mental abilities. Although these thought processes will not be discussed often, a major objective of most engineering courses is to contribute to the development of your thought skills. The fact that these processes are seldom treated explicitly can be misleading; educators and employers do not underrate their importance.

Do not underestimate, as many embryo engineers do, the value of skill in verbal communication. Eventually, on-the-job experience will convince you, if advanced warnings have failed, that this is a crucial skill. But why not follow the advice of graduates and employers by developing your communication skills during your college years?*

Surely you know that you also must be skilled mathematically. It is an important skill that is appropriately emphasized in the chapters on modeling. It will be needed most in modeling alternative solutions in order to predict their behavior.

Graphical skills provide an important means of expression, especially in engineering, where drawings, sketches, and graphs are so useful. Graphs, for example, are often used to communicate concepts, to aid thinking, and to detect as well as describe relationships between variables. The ability and inclination to think and communicate graphically are developed throughout an engineering education.

Naturally you will rely extensively on your computational skills, especially the ordinary pencil-and-paper type that you know so well. But many computational tasks in the real world are so lengthy that faster computational means, such as computers, are necessary. Therefore, in college, you are equipped with an array of computational tools suitable for jobs ranging from simple one-shot calculations to computational tasks involving millions of repetitive operations.

Another skill, the need for which became apparent in Chapter 1, is the ability to work effectively with other people. The day-to-day practice of engineering brings you in contact with many people in many capacities. Inability to maintain cooperative working relationships with at least most of them spells trouble.

So much for the skills you will need. And here's good news:

*I recall such exhortations from my college days, and I remember how vaguely I grasped the importance of what my teachers were telling me. Now, however, I can testify to their wisdom!

they are stressed in an engineering education. It is good news for two reasons. First, such thought skills are virtually invulnerable to obsolescence—most of them are good for a lifetime. Second, these skills are generally useful in other careers (e.g., in management). Skills in problem solving, communication, mathematics, and the like are widely transferable.

Knowledge

You must also possess a wealth of factual knowledge in areas such as science, engineering systems, and economics if you are to design complicated machines and structures. In Part 3 you will examine samples of this knowledge in detail, and this will give you further insight into engineering *and* information that you will subsequently find useful.

Attitudes

Certain qualities you will bring to bear on problems involve neither factual knowledge nor skills. They constitute what can be described as attitudes. They are difficult to define, but certainly include a questioning attitude. This curiosity for the "how" and the "why" of things leads to much useful information and many useful ideas. Some of this questioning results from inquisitiveness, but some arises from a certain skepticism that prompts the engineer to challenge various "facts," requirements, and features to make them "prove themselves."

In such a dynamic field, flexibility—receptivity to changing concepts, innovations in technique, and new ideas—is obviously desirable. Furthermore, working in a dynamic field has another and related consequence: a favorable attitude toward learning after graduation is of utmost importance. This calls for a certain humility—recognition that a college diploma symbolizes the end of the beginning of a lifelong process. The willingness to continue learning is a major determiner of success in any profession.

Clearly, however, your attitude toward people—your employers, colleagues, and especially the people directly and indirectly affected by your solutions—is the most important factor. This attitude should be consistent with what is traditionally expected of professionals. The professional person serves society as an expert with respect to a type of relatively complicated problem. The layman trusts the professional person and, because of this, the professional has an obligation to perform his services ethically. Since most of your creations will affect the well-being of many people, the public trusts that your designs will be safe and otherwise beneficial to the welfare of mankind. The public also trusts that it will receive the full measure of service for which it pays.

Your professional obligations involve more than living up to this trust. They include:

- **Insistence on seeing a project through to successful implementation of a solution.**
- **A desire to follow up on that solution in order to benefit from experience with it.**
- **A feeling of responsibility to your colleagues expressed in your actions, your attempts to improve the status of the profession, and a willingness to exchange information.**
- **Maintaining in strictest confidence the unpatented ideas, secret processes, and unique know-how that your employer or client considers proprietary.**
- **A concern for the direct and indirect consequences of your solutions.**

These are no small matters.

Granted there are people with no special training who are clever enough to design tools and other devices (but not long-span bridges or communications satellites). Granted, too, that their creations might suffice as long as they are not to be mass produced, do not see heavy use, and do not jeopardize life and limb, and as long as no one insists on high reliability at low cost. Almost anyone could develop a method of fabricating reed switches, although contact alignment and spacing would be a challenge. But believe me—there is a big difference between putting a few devices together at a workbench with a bunsen burner and some hand tools, and manufacturing millions of devices where thousandths of a cent and thousandths of an inch count! Not that many people can create complex physical contrivances for society *and* meet the demands for low cost, high reliability, long life, large volume, and safety. Those who can are engineers. To perform their function effectively, engineers must possess certain qualities that align rather nicely into three major categories: *skills* (for instance, design, measurement, and communication), *knowledge* (of physics, history, economics, feedback systems, etc.), and *attitudes* (objectivity, social concern, willingness to continue learning, and the like). In my opinion, this pattern is important. The remainder of this book elaborates on certain of those skills, types of knowledge, and attitudes.

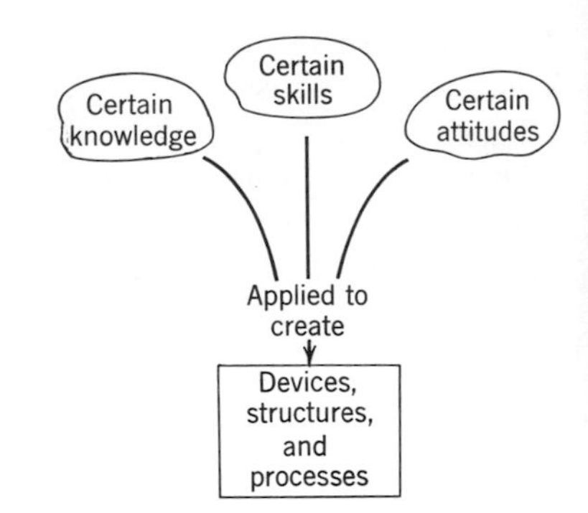

If you are discouraged by what may seem like an imposing set of qualities, do not overlook the following. Note that this chapter characterizes a *good* engineer and, moreover, a good engineer some years after graduation. These qualities come partly from college education, and partly from experience as a practicing engineer. The prime objective of an engineering program is to provide you with a substantial start in developing these qualities. Keep in mind, too, that a student rarely fails to graduate from an engineering program because of lack of ability.

Returning to the chapter-title question, "What Does It Take?", the answer in a nutshell is, "It takes a lot." However, engineering *offers* a lot and, since we have determined what it demands, "equal time" should be given to the rewards, challenges, benefits, and other pluses of an engineering career.

EXERCISES

1. In almost every instance the engineer must propose a solution to a problem in a rather limited period of time. What do you imagine are the consequences of this restriction? (For example, what must he resort to, what must he forego?)

2. Take a familiar device or structure and attempt to identify some of the conflicting objectives with which the designer presumably had to cope as he arrived at that solution. (For example, in most cases, he has to resolve this conflict: maximize the number and effectiveness of functions the device performs but minimize the cost of manufacturing it.) For this purpose you might select a camera, a household appliance, or a campus building.

3. In this chapter there is a statement to the effect that the practice of engineering is more of an art than you may have assumed. Looking through Chapters 1 and 2, what circumstances make it partly an art?

4. Many engineering projects require the combined efforts of two or more engineers, sometimes dozens or even hundreds. Identify some of the benefits as well as adverse consequences of such group efforts for the engineer (in the short and long run), for the enterprise, and for the solutions such teams create. You will find some clues in Chapters 1 and 2.

WHAT DOES IT OFFER? Chapter 3

What, besides a respectable living, does an engineering career offer? Every individual answers this question differently, but they all agree that it offers certain rewards. The main one is the satisfaction derived from *creating*. Individuals have their own ways of expressing it: "something tangible results" . . . "a welcome outlet for my long pent-up imagination" . . . "a thrill known also to the artist" . . . "visible solutions to real problems" . . . "the feeling is a good one." Each, in his own words, is telling about the joy that comes from creating useful structures, machines, and other complex physical contrivances.

Furthermore, this profession offers variety, which is apparent in Chapter 1. There is variety *within* each project; day-to-day activities include thinking, meeting, drawing, calculating, telephoning, corresponding, experimenting, searching in the library, consulting colleagues, and talking to customers. Similarly, there is variety from assignment to assignment. One project may involve designing a household water purifier, the next might be a fact-gathering assignment, the next might involve troubleshooting of a product already on the market, and the next the design of a water desalinization plant for a medium-sized city. Finally, there is variety in job opportunities.

OPPORTUNITIES

Certainly one factor that makes an engineering education attractive is the uniquely broad range of career options open to graduates. There are more sides to this profession than you have surmised; many types of work, many fields of specialization, and many problem areas. There are possibilities here to suit a broad range of talents and interests. One word sums it up: *variety*.

Many Specialties

There are the long-established, major branches of engineering, such as aeronautical, chemical, civil, electrical, industrial, mechanical, and metallurgical engineering. In addition, a number of new ones have evolved recently, such as those devoted to space travel and medical instrumentation. Furthermore, in practice, most engineers concentrate their efforts on one phase of a major specialty. For example, some mechanical engineers are experts in the design of mechanisms, others concentrate on refrigeration, and another group deals in propulsion systems. Thus, considering all the branches of engineering and the numerous subdivisions of each, it is apparent that there is a wide variety of specialties from which to choose.

Many Types of Activity

Within any specialty there is a broad range of activities you can choose from, defined by the following extremes. At one end of this spectrum is development engineering, the frontier type of job; the problems generally call for novel or technically advanced solutions, as in VTOL development work. This type of activity is demanding. You must be well trained in science, mathematics, and advanced engineering, and you must keep up to date in these areas. Engineers who are happy and successful in this type of job thrive on a steady diet of complex technical problems.

In contrast to this is what is commonly referred to as sales engineering, involving a minimum of technical innovation but a maximum of involvement with people. Superior salesmanship, a keen interest in people, and a pleasing personality are highly desirable for this type of activity.

Between these extremes there are hundreds of jobs that differ in day-to-day activities, challenges, and technical demands. They suit a broad range of abilities and preferences. Take your choice. All jobs over this broad spectrum involve problem solving; all involve working with things and with people; all require salesmanship; and all require creativeness and the other qualities described in Chapter 2, but in varying degrees.

Many Types of Enterprises

Among the traditionally heavy employers of engineers are the aircraft, appliance, automobile, chemical, communications, construction, electronics, energy, metals, machinery, and transportation industries. The tempo of innovation in products, services, and manufacturing processes is accelerating, so that the demand for engineers in these industries continues to increase. This is true, by the way, even of the technical enterprises that you may feel have long since become static. Consider bridge design and construction, which could be cut and dried and therefore a relatively dull kind of engineering activity. It is quite the contrary. Innovation is conspicuous in this field (Figure 3-1); imaginative efforts to reduce cost, enhance appearance, and provide variety could provide the bases for a fascinating book. The point of this is that industries you might write-off because you assume they are technically stagnant, in most instances do offer many challenging opportunities. Then there are the newer "glamour" industries (e.g., those associated with space and computers) that have become major employers of technical people.

Other Opportunities

True, a majority of engineers are employed at all levels by industrial firms, public utilities, large consulting firms, private contractors, and government agencies. But that does not mean there are no opportunities to go in business for yourself or to work in a small organization. There are self-employed engineers, and even more of them are involved in partnerships, usually specializing in some form of technical consulting.

Of course, not all engineering graduates practice engineering. There are more scientists, physicians, lawyers, architects, teachers, and particularly business executives who hold undergraduate engineering degrees than you probably realize. The numerous career paths that lead from an engineering education are worth noting. When it comes to the number of career options at graduation, there isn't another college degree that can equal a bachelor of science in engineering.

CHALLENGE

Besides rewards and variety, engineering offers *challenge*. To help you envision what that might amount to for you, here is a sampling of problem areas that will be challenging engineers over the next several decades.

Figure 3-1a *This unique bridge across the Severn River in Great Britain is a significant departure from the familiar suspension bridge. The deck and what for a conventional suspension bridge would be its supporting truss are an integral unit, forming one large, continuous, hollow beam. The cross-sectional shape of this box beam, apparent in the photograph, is aerodynamically superior to the conventional rectangular truss. It has reduced the effects of cross winds by two-thirds and, as a consequence, significantly reduced the steel required throughout the structure. Each section of the deck, like that shown being hoisted into position, is fabricated on shore and then floated to the site. Adjacent sections are welded together to form one 990-meter beam, which becomes the main span. These and other innovations in design and construction methods resulted in a bridge costing about one-fourth as much as a conventional suspension bridge of comparable size.*

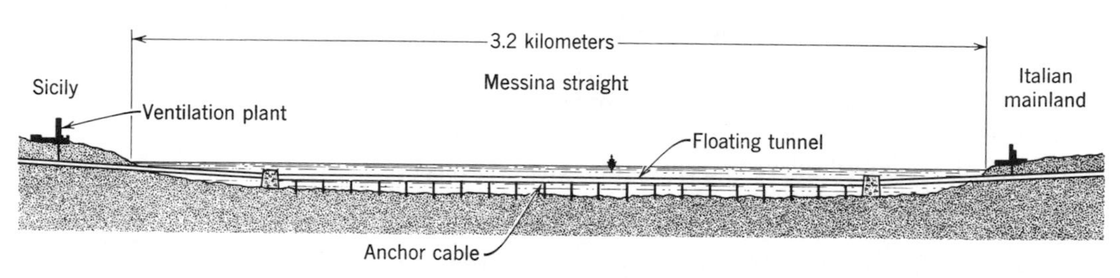

Figure 3-1*b* *A bridge is not the only type of fixed crossing. In addition to the tube-in-trench concept employed in the Chesapeake Bay Bridge-Tunnel, there is the bored tunnel and the innovation modeled here—a floating tunnel (floating bridge, if you prefer). This new type of underwater crossing, proposed to connect Sicily to the Italian mainland, is especially attractive since the strait is about 3.2 kilometers wide—a little too wide for a suspension bridge. Furthermore, shipping traffic is such that a bridge is undesirable, the water is so deep that a bored tunnel is impractical, and frequent earth tremors in that area make bridge and bored tunnel unsafe. This floating tube presents no hazard to shipping, it is 46 meters below sea level, and it is relatively unaffected by seismic shocks.*

Natural Resources

Many of the career opportunities, much of the excitement, and a large share of the challenges awaiting young engineers in some way relate to the matter of resources. That covers a lot, including human, financial, information, material, food, water, and energy resources. For any one of these resources, the engineer may be involved in discovering sources, developing conversion methods, improving utilization, or developing means of recovering the expended resource.

Energy As a Case in Point

Surely you are generally aware of the rising demand, shrinking reserves, and other gloomy dimensions of man's energy problems. But you may not be so familiar with the variety of fronts on which engineers can contribute solutions and, therefore, where you might fit in.

SOURCES. The basic sources of energy have been known for a long time, so nothing can be done about developing new sources. But the story is quite different in a closely related area. There will be a lot of activity in the search for new *supplies* for known sources of energy (e.g., new oil, coal, and uranium deposits). This includes the development of more effective means of locating underground and underwater deposits, as well as the search itself.

Furthermore, you may have observed that what will shortly be depleted for a number of resources are not the supplies but the *readily accessible* supplies. We are being forced to turn to difficult and expensive deposits of oil in the Arctic, oil mixed with shale, and oil beneath thousands of feet of ocean water. For instance, we have known about vast deposits of oil mixed with sand for centuries, but we never bothered with them—it wasn't necessary. Now we are interested. The development of technically, economically, and environmentally satisfactory systems for recovering oil under harsh conditions like those encountered at sea are better known and even more challenging (Figure 3-2*a*).

CONVERSION. There are a number of known but, thus far, virtually useless sources of energy. What's desperately needed are economical methods of converting the energy potential of

Figure 3-2a *The sea is one of the "new frontiers" for engineers, offering an impressive variety of challenges. Among them are: development of new types of ocean-going vessels, such as giant air-cushion freighters capable of 100 km/hr speeds; development of systems for the recovery of important minerals from the ocean bottom; and the the design and construction of oil-drilling platforms like this rig shown headed for the turbulent North Sea. This type, known as a semisubmersible, has a platform area of 2800 square meters and houses 50 workers. It will be towed to the drill site, partially submerged, and then anchored until its work is done there; then it will be refloated and moved to the next site.*

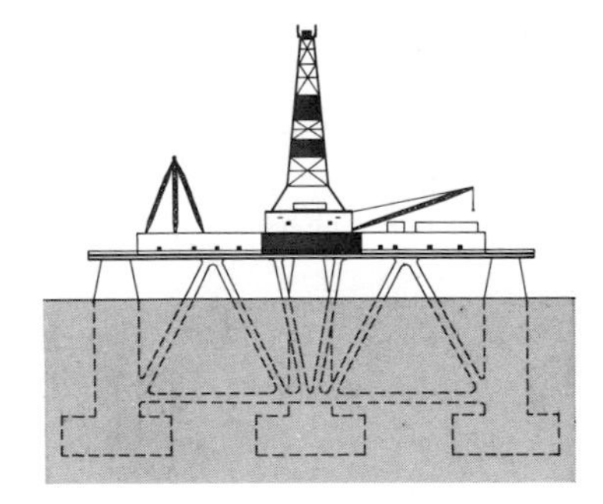

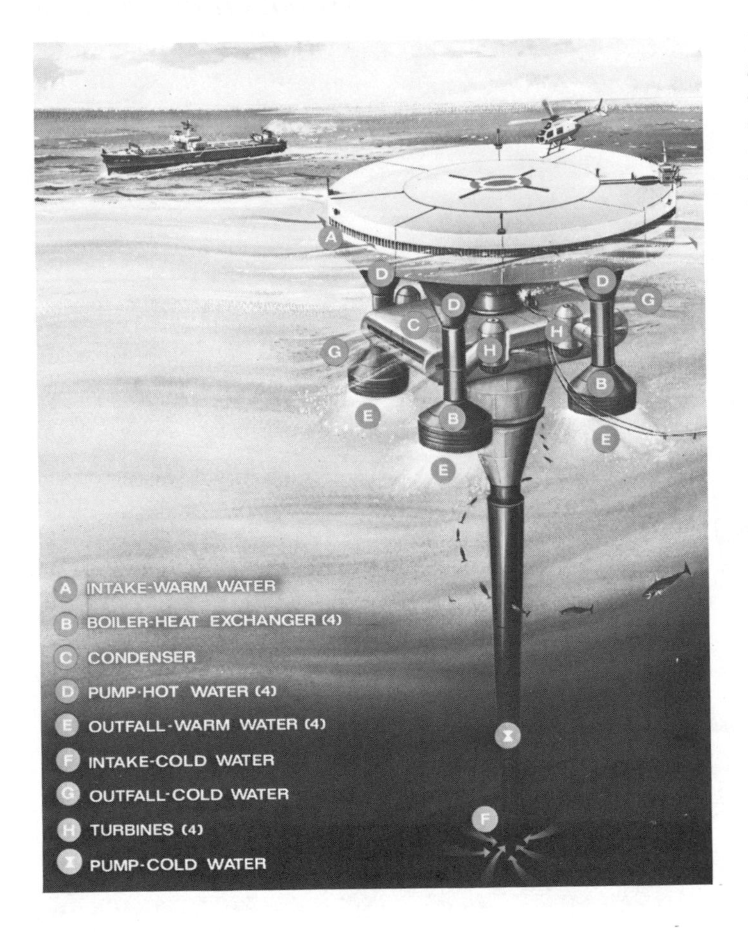

Figure 3-2*b* *Much of the sun's energy reaching the earth is absorbed by the oceans. One scheme for tapping this energy closely parallels that employed by the steam turbine used in power plants. The main difference is that the "working fluid" is ammonia instead of water. Fortunately, ammonia vaporizes at the temperature of surface water and condenses at the temperature of deep water. The low-pressure ammonia gas propels a turbine, which drives an electrical generator. The accompanying diagram traces the complete cycle. The enormous platform pictured here houses the necessary boiler, turbines, condensers, and pumps required for this process. It would require very little human attention for ordinary operation. Such energy platforms are still in the conceptual stage; there are numerous problems to be resolved, but the potential and promise are attractive, indeed.*

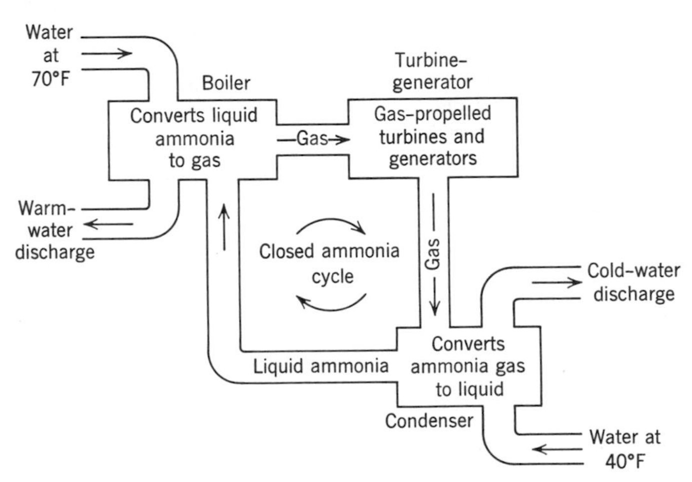

Figure 3-2c *Energy, minerals, and food are not the only resources man can extract from the oceans; fresh water is available, and not only through desalinization. Several recent feasibility studies indicate that it is technically possible and economically attractive to move icebergs from the polar regions to arid areas of the world to provide fresh water. This would hardly dent the supply of polar ice–approximately three-fourths of the earth's fresh water is ''locked'' in the ice caps. An iceberg of 3 million cubic meters would be capable of irrigating thousands of square kilometers of cropland. However, some major problems remain; for example, how do you harness an iceberg if you intend to tow it? Yet those who have studied the prospects are optimistic that someday we will see iceberg trains passing on their way to market.*

Figure 3-2 d *A city on the sea? A number of detailed proposals for exactly that have emerged over the last decade, and herein lies one of the most exciting and most controversial prospects on this new frontier. Most designs call for a floating platform employing the same principle apparent in Figures 3-2*a, *3-2*b, *and 3-2*e. *All employ slender, vertical structural members that are relatively unaffected by waves and swells. Floating platforms built on this principle can be as large as purpose and economic feasibility dictate, such as a jetport. A city on the sea could be virtually self-sufficient in terms of energy, water, food, and waste management. Furthermore, they are movable, which brings to mind the prospect of migration–floating with the season. (Wouldn't that confuse the mailman?)*

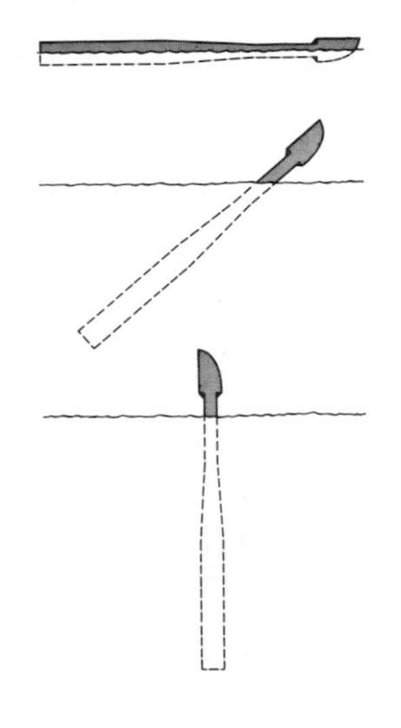

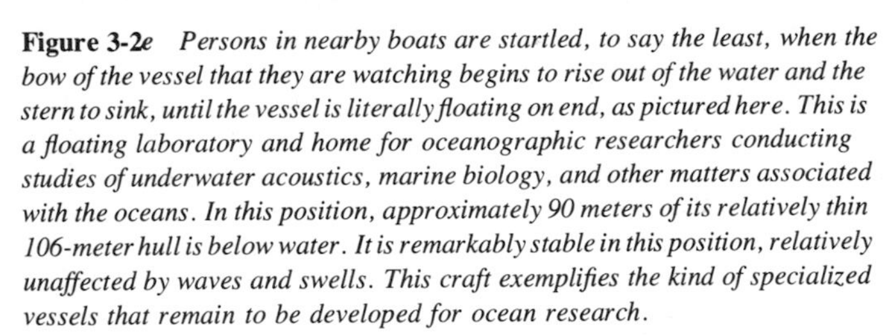

Figure 3-2e *Persons in nearby boats are startled, to say the least, when the bow of the vessel that they are watching begins to rise out of the water and the stern to sink, until the vessel is literally floating on end, as pictured here. This is a floating laboratory and home for oceanographic researchers conducting studies of underwater acoustics, marine biology, and other matters associated with the oceans. In this position, approximately 90 meters of its relatively thin 106-meter hull is below water. It is remarkably stable in this position, relatively unaffected by waves and swells. This craft exemplifies the kind of specialized vessels that remain to be developed for ocean research.*

these sources into useful forms. This is where a lot of energy research and development funds will be applied and, therefore, where much of the action and many of the jobs will be. Large-scale, economic methods of converting the energy potential of sunlight, wind, and refuse are in various stages of development; some are near success (Figure 3-3) and others are far from it.

Figure 3-3 *You might have trouble recognizing it as such, but it's a windmill at the mouth of a large wind tunnel. This is a vertical axis type, with blades of balsa wood coated with fiberglass. It is not only simple and cheap, it is multidirectional (horizontal axis designs have a problem in this respect). This windmill, producing enough power to supply a typical household, is one of a variety of designs under development around the world.*

Here is an excellent case in point. The theory of generating electricity by an ingenious method that capitalizes on the temperature differential between surface water and the depths of the ocean goes back almost a century (Figure 3-2*b*). It's a beautifully simple concept, appealing indeed. But can it be developed into a practical conversion scheme? (Notice that, since it is sunlight that causes this differential, this is another way of capitalizing on the energy of the sun.)

STORAGE. Has it occurred to you how ill-equipped we are to capture and *store* energy from intermittent sources like sunlight, wind, and the tides? A breakthrough in energy storage capability would be of enormous technical, economic, and social significance. True, we have batteries, artificial lakes, molten salts, flywheels (Figure 3-4), and numerous other possibilities but, as of now, we are lacking a high-capacity, low-cost storage medium.

Such a capability offers more than the obvious payoff. We could obtain much better utilization of our power generating plants if electricity generated during the low-demand periods (the middle of the night, for example) could be stored and fed back into the system during peak-demand periods. The storage breakthrough we seek would produce the same effect as a 25 percent or more increase in power-generating capacity—nothing to sneeze at.

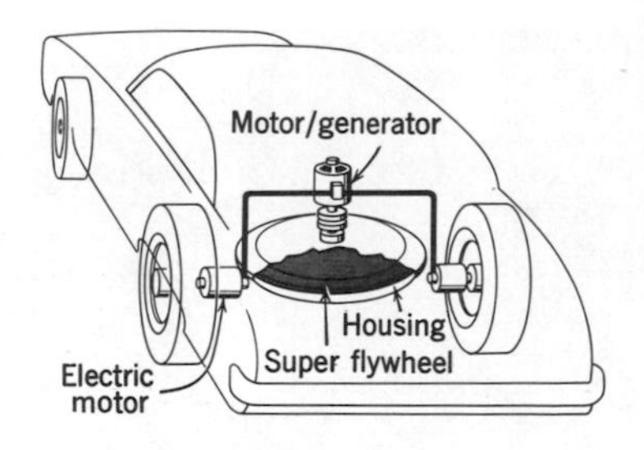

Figure 3-4 *It is practical to propel a vehicle using energy stored in a rapidly spinning flywheel. This automobile employs one flywheel, a motor-generator, and two small drive motors. The motor-generator functions as a motor to spin the flywheel during charging. When the car is in motion, the flywheel drives the motor-generator which, functioning as a generator, produces electricity to drive the wheel motors. One of the appealing features of this scheme is that when the driver applies the brakes, that force is transferred to the flywheel, so that what goes to waste as heat in standard vehicles is captured and reused. Systems like this will propel a bus for about 32 kilometers on a four-minute charge, while a small automobile can function for about two hours.*

This is possible primarily as a result of recent developments in high-strength materials and low-friction bearings. These, along with a vacuum spin chamber, make possible "super flywheels" with very high spin speeds and run-down times on the order of months.

UTILIZATION. Much of the work of an engineer involves resource allocation, and so he is a major influencer of the rate at which man consumes a resource like energy. How could it be otherwise? The designer of a building determines the type and amount of energy the building and its occupants will consume over many decades. The designer of a steel mill does likewise, and so do designers of transportation facilities, communication systems, and agricultural equipment. Engineers determine not only what kind and how much, but the amounts of energy lost through friction, leakage, and the like, and how much "waste energy" is reclaimed and profitably used.

The growing pressure to conserve energy has added a new dimension to the work of all engineers whose creations consume energy. In many instances, like automobile design, it's a new era compared to the days when oil was considered abundant. To cite just one case, the designers of automobiles, trucks, and trains, from now on, will be expected to devote careful attention to the aerodynamic performance of their vehicles. Wind resistance significantly affects energy consumption, as Figure 3-5 illustrates. These added considerations in design mean new challenges and increased demand for technical talent.

A Similar Story for Material Resources

You don't read about it as much as energy and food problems, but people, particularly in the metals industries, have come to recognize material resources for what they are—limited. This has important consequences for engineers, because they have a lot to

Figure 3-5 *This simple sheet metal "picture frame" attached to the leading face of a truck body illustrates how some attention to the aerodynamics of vehicle design can pay off. Trucks equipped with this inexpensive device have been found to use 23 percent less fuel than similar standard trucks at 90 kilometers per hour speeds. If all medium and heavy trucks were equipped with such devices, the fuel savings per year would be in the millions of barrels.*

say about what materials will be used, in what quantities, with how much waste, with how long a useful life, and with what possibilities for recycling. Of course, it has always been their objective to minimize the material resources consumed by their creations without jeopardizing the users' safety. This is an important part of their job—to labor long and hard to "do the best with the least." A good example of how engineers can substitute ingenuity for materials is the clever scheme they devised to conserve the copper that goes into telephone wire. It is a device that enables a single wire to carry more than one telephone conversation simultaneously. Where the cost would be especially high, such as with transoceanic submarine cable, one strand carries more than a hundred conversations. You can imagine how many wires would be strung around the countryside if a separate strand were required between each two parties! Observe that engineers influenced the amount of a resource consumed —copper—through their design. Hopefully this will always be true, with even greater attention, care, and ingenuity in the future.

There are parallels between the problem areas associated with energy and material resources, with one notable exception. The opportunities for recapturing spent resources are excellent for materials, but negligible for energy. You are aware of the recycling process and its desirability, but you may not fully realize just how much the economic and technical feasibility of recycling are influenced by the *designer* of the product. This gives rise to a major challenge: to significantly increase the proportion of our natural resources that can be reused through careful forethought during design. To be sure, designing recyclability into products adds another dimension to design.

Other Resources Too

Don't let the relatively few words devoted to resources other than energy and material resources mislead you. Elaborating on all the high-opportunity areas is impossible. Energy and materials were chosen as representative examples of what is true for other resource problems. You read the newspapers, so you are aware of food and water shortages. You can take it for granted that a lot of imagination and plenty of technical manpower will be needed to alleviate such problems. Figures 3-6 and 3-7 indicate what has been accomplished. As an illustration of what might be done, consider the possibility that man may eventually embark on multibillion-dollar construction projects for redistributing fresh water over whole continents. (This has been proposed for North America and calls for distribution of Arctic fresh water through Canada, the United States, and Mexico in order to meet their requirements.) Although they may strike you as desperation measures, these schemes and the one pictured in Figure 3-2*c* may come to pass; the problems are that serious.

Solid-Waste Management

This is one aspect of the broader matter of resource management and another problem area that will provide many future jobs, in two respects. One is in *developing* new collection, sorting, recycling, and disposal methods and means of generating energy from material waste. The other is in *applying* these systems on a community-by-community basis. This is where most of the jobs in the waste management field will be found.

Environment-Related Opportunities

There is a parallel between this area of opportunity and solid-waste management in that (1) you are generally aware of the nature and magnitude of the problems, (2) there is need for imaginative technical talent to develop new pollution abatement and control capabilities, and (3) the majority of jobs will be not in development but in application of abatement and control systems to steel mills, power plants, refineries, and other sources of pollution.

Potential Contributions in the Health-Care Field

A sampling of recent achievements indicates what you might contribute in the future. The modern operating room is filled with impressive-looking machines that are routinely employed. There also are accomplishments such as the heart-lung machine, the miniature devices for incisionless internal surgery, and the equipment for neurosurgery by freezing.

For medical diagnosis there are tiny instruments that can be inserted into vital organs (including the heart through a vein) to make measurements, apparatus that automatically analyzes blood samples, and computerized analysis of electrocardiograms. Other diagnostic innovations are in the developmental stage.

Prosthetic devices are numerous, indeed. There is the miniature battery-operated heart pacer that is embedded in the patient's body to keep his heart operating by sending electrical pulses directly into the heart muscles. A number of artificial organs and vastly improved artificial limbs are under development. The most dramatic are the artificial kidney and the mechanical heart.

Exciting developments are taking place in hospitals. A variety of ingenious sensors are being developed for continuous measurement of patients' temperature, blood pressure, and so forth, so that these vital signs can be monitored and an alarm sounded when a patient's condition warrants attention.

In spite of these developments, some major challenges remain. Here are some of them.

- **Since hospitals are so highly instrumented, they need "resident engineers" to oversee the acquisition, installation, and care of this complicated equipment, and particularly, to protect the safety of persons coming in contact with it, in view of the high voltages or radioactive materials that are often employed. All hospitals need such technical advisors, but few have them.**

(a)

(b)

Figure 3-6 *Without some help you might be hard put to explain the disks pictured in* (a). *These are fields of alfalfa, barley, and wheat, each about one million square meters in area, made possible by irrigation machines like the one pictured in* (b) *and* (c). *Each machine pumps water from a well at the center of the circular field and distributes it through a pipe that sweeps as a radius every 24 hours. The 560-meter pipe is supported by a superstructure mounted on rubber-tired carriages spaced about 60 meters apart and driven by electric motors. (The speeds of these motors must be precisely synchronized so that adjacent sections remain aligned as the whole thing rotates. That is a challenge.) A series of such machines is gradually converting formerly useless African desert into croplands–presenting an odd but most welcome sight.*

Figure 3-7 *These experimental, air-supported greenhouses, located on the California Peninsula, are forerunners of what will probably be a revolution in agriculture. They make it possible to grow fresh vegetables and other crops in an arid climate. In this installation, seawater is desalted by distillation employing the waste heat from diesel engines used to drive the electric generators. The desalted water, with small amounts of chemical fertilizers added, is distributed via a trickle system to support growth. The plastic greenhouses significantly reduce the water that would otherwise be lost to the atmosphere and permit close control of temperature and humidity. A fullscale, commercial version of this method is now in operation at an extremely dry site on the Arabian Peninsula, with 48 air-supported greenhouses. The closed-environment production of food has enormous potential–besides conserving water, it is climate-independent, eliminates the adverse environmental damage now done by fertilizers and pesticides, increases the tillable land area, and improves crop yields. There is lots of room for improvements in this process, which is where you might come in by applying solar energy and closed-cycle operation. Developments like this and the desert irrigator (Figure 3-6) should be an inspiration to those who are inclined to devote their technical talents to mankind's food and water problems.*

- **There is a need for the continued development of diagnostic aids (specialized instruments and computer programs) that will make diagnostic services available at low cost to masses of people (Figure 3-8). Machines that quickly and cheaply analyze test specimens, electrocardiograms, X-rays, and other diagnostic evidence will significantly advance the prevention and early detection efforts of physicians.**
- **The information-handling problems of a hospital are approaching the impossible. Information handling includes gathering, storing, distributing, processing, and retrieving information on patients. You wouldn't**

believe the cost of all this. Nor would you like to hear about all the misdirected, misplaced, transposed, and too-late information, nor the gruesome consequences of these slipups. The computer will be a boon in these respects, but people are needed to do the applying which, on the basis of attempts so far, appears to be especially challenging.

- **There is the perennial problem of making inventions widely available at reasonable cost. For every inventor of an artificial kidney, hundreds of engineers are needed who are willing to refine the original and usually very costly machine in order to reduce the cost to a tolerable level. This requires persistent, imaginative technical effort, which is absolutely essential if a significant number of people are to benefit from the basic idea.**

This sampling gives you an idea of the engineering opportunities available in the field of health care. Incidentally, there are some established specialties within the field of "medical engineering." The *biomedical engineer,* for instance, focuses on the development of instrumentation for medical research, diagnosis, treatment, care, prosthesis, and education (Figure 3-9). There are a number of degree programs in biomedical engineering. The *clinical engineer* is also concerned with instrumentation, prosthesis, and the like, but on a patient-by-patient basis. He works as a member of a medical team, supplying the expertise needed when, say, a special device must be developed to be implanted in a particular patient. The *hospital engineer* is a technical advisor and overseer, whose responsibilities were outlined earlier. Notice that although there are some emerging subspecialties in the field of medical engineering, your opportunities in the field are certainly not restricted to those labeled subspecialties.

In Sum . . .

Because natural resource and health care problem areas have been emphasized do not conclude that "that's it." These illustrate in some depth what is to be found in many problem areas.

An analysis of the problem areas in which the engineering profession stands to make significant contributions reveals a

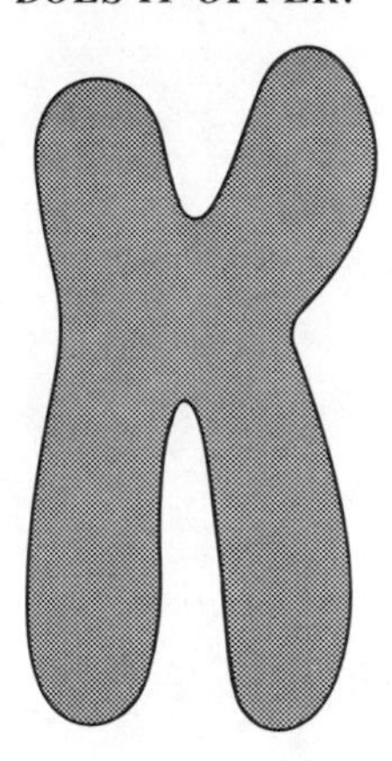

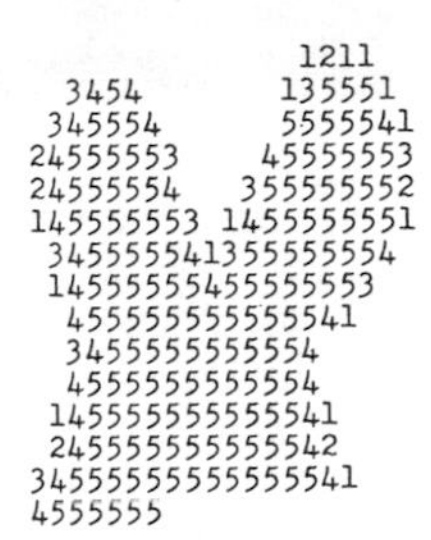

Figure 3-8 *In medical research and diagnostic work endless hours are spent analyzing photomicrographs of chromosomes. In one such photograph there are 46 chromosomes; extracting useful information from the photograph requires a lot of tedious work. But special machines may eventually come to the rescue. One sophisticated piece of apparatus already developed can convert chromosome shapes in a diagnostic photograph into numbers in a fraction of a second. In the process, called digitizing, the machine scans a shape like the one shown and converts what it sees to numbers representing shades of grey on a scale of 1 through 5. Shown here is the digitized version part way through the conversion scan. This is a significant development because photomicrographs and X-ray photographs in digitized form can be processed by computers, more rapidly and more cheaply than by present methods.*

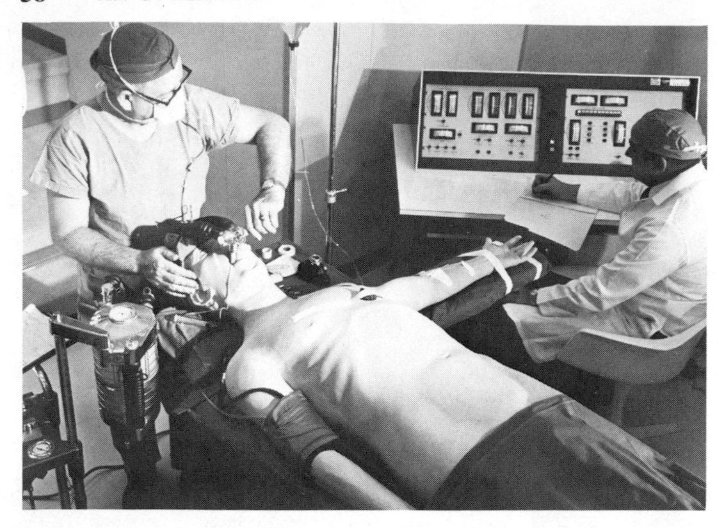

Figure 3-9 *This is a computer-controlled patient simulator for training anesthesiologists, developed by engineers. It "breathes, coughs, and vomits," has a "heartbeat," and can "die." In these and many other respects Sim One closely mimics human behavior. The student can give injections, administer oxygen, check pulse rate and pupil dilation, and perform most other routine and emergency functions of an anesthesiologist. The instructor, using the console at the right, can program the lesson, evaluate performance, temporarily stop the process to discuss a point with the student, and have a lesson repeated if performance is not satisfactory. Obviously, these are privileges not available when real subjects are used. This simulator significantly shortens the training period for an anesthesiologist and allows instructors to exert much closer control over the learning process.*

special need for engineers who offer a *broader perspective,* a *greater imagination,* and a *higher sense of purpose* than have prevailed up to now. On the matter of broader perspective, interrelated problems are too often attacked separately, as is the case with the many interdependent problems that plague a city. Often a solution to one problem aggravates other problems. Engineers are needed who will view a situation with broader perspective than has been customary and produce a well-integrated solution to the whole problem.

There is need also for engineers who will bring fresh ideas to the many long-standing, still unsatisfactorily solved problems. Society can use engineers who are bent on *imaginative* application of their knowledge and skills.

Engineering also needs many more practitioners dedicated to applying their talents where society's needs are greatest. As an engineer you can contribute toward easing man's resource plight, improving the environment, extending low-cost medical care, combating starvation, alleviating conditions that breed social ferment, aiding the physically disabled, improving travel safety and mass transportation, converting wastelands into croplands, and aiding underprivileged nations.

EXERCISES

1. Write a short paper (or prepare a brief talk) summarizing engineering's potential contributions in one of these areas: surgery; prosthetics; air safety; urban mass transportation; energy storage; food resources; crime prevention and detection; pollution abatement and control; education; fire prevention and control; or you or your instructor pick an area.

2. It would be enlightening, perhaps even helpful to your career planning, to have an alternatives tree for the many career possibilities that can evolve from an undergraduate engineering education. Construct one.

Skills

PART 2

Some people call engineers "problem solvers extraordinary." This is an exaggeration, but they *do* have a knack for solving problems. It comes about mainly from at least four years of intensive drill in doing exactly that. Recognize—starting now—that problem solving is the very heart of engineering and, furthermore, that problem solving, modeling, and optimization—the subjects of Part 2—are *skills*. Since skills require practice and therefore time, it pays to start the development process now.

DESIGN: DEFINING THE PROBLEM

Chapter 4

Do you advocate trying to solve a problem without knowing what the problem is? Surely you *say* no, yet you probably *do* exactly that in practice. In casual problem solving this habit matters little but, when problem solving is your business, it can be consequential, indeed. Thus, as an engineer, it is important that you clearly define a problem *before* you become involved with solutions.

To do that, however, requires that you have a more specific notion of what constitutes a problem. Chances are you have never given that word much thought, and why should you? In conversational use a strict definition is hardly necessary; everyone knows more or less what the word means, and that suffices. But it does not suffice for an engineer attempting to characterize a specific problem prior to solving it. *Then* a fuzzy idea of what constitutes a problem is a handicap. Therefore, to satisfactorily solve problems when you are being paid to do so, you should have something specific in mind.

What Constitutes a Problem?

A problem arises from a desire to achieve a transformation from one state of affairs to another. These states might be two locations, the interval between which must be traversed (two points on opposite sides of a river, one city and another, Earth and another planet). Then, too, many problems involve a transformation from one form or condition to another (e.g., brackish to potable water). In every case there is an originating state of affairs, call it ''state A'', and a state of affairs the problem solver seeks a means of achieving, call it ''state B.'' Figure 4-1 provides examples and so does Chapter 1: a ''means of *transforming* patient information into a set of ailment probabilities,'' a ''means of *transforming* glass tubing, reeds, and gas into the specified switches.''

A *solution* is a means of achieving the desired transformation. It is difficult to envision a problem to which there is only one solution; for most problems there are more alternative solutions than there is time to investigate. Think of the many modes of travel and all the possible routes they can be combined with to provide alternative means of getting from one point on the map to another.

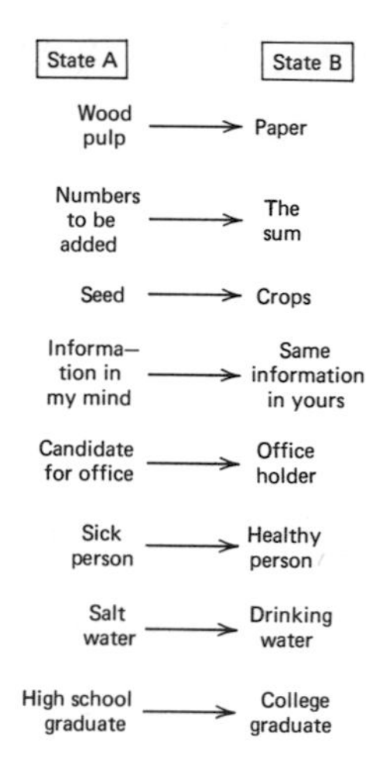

Figure 4-1

But that's not all; a problem involves more than finding *a* solution; it requires finding a preferred means of achieving the desired transformation (the mode of transportation that is best with respect to cost, speed, safety, comfort, and reliability). A basis of preference among solutions is commonly referred to as a *criterion*. Some criteria employed in the VTOL problem (Chapter 1) are the plane's cost, safety, reliability, noise level, maneuverability, and speed. Engineers are accustomed to thinking about, evaluating, and struggling with criteria.

They are also accustomed to living with another common characteristic of problems: *restrictions*. A restriction is an inescapable solution characteristic—a ''must'' for a solution if it is to be eligible for consideration. Chapter 1 provides examples, like ''must produce results within one minute.'' You can spot others in those case histories.

There are still other problem characteristics you must cope with but, by now, it should be apparent that as a professional problem solver you must have specific characteristics in mind when thinking about a problem. It follows, too, that you had better have an effective method of solving problems. To help you develop one, here is a short discourse on problem solving.

Problem Solving

What NOT to Do

The management of a company that distributes livestock feed is concerned about the high cost of handling and storing its products. At present the materials are bagged and stored by the procedure diagrammed in Figure 4-2. An engineer is assigned to find a less costly method.

A common tendency is immediately to begin thinking of improvements in the present solution. In this problem, you must

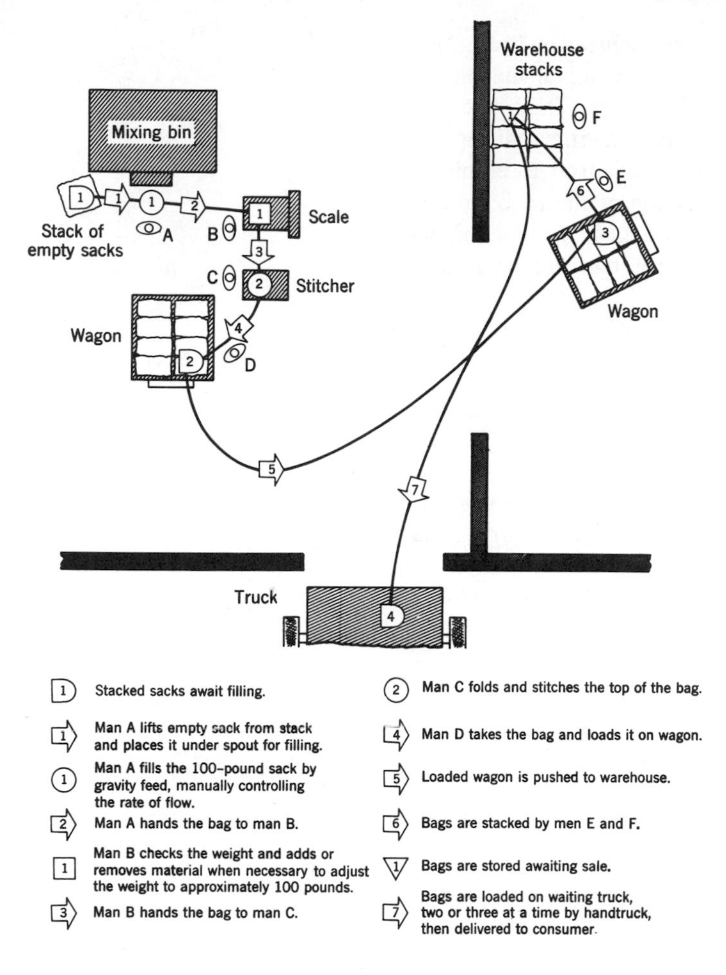

Figure 4-2 *The present method of filling, storing, and loading sacks of feed.*

admit, it is tempting to begin by scrutinizing the solution described in Figure 4-2, seeking improvements that might reduce the cost of the process. If you were to do so, you would become concerned with matters such as equipment for filling, weighing, and sewing the sacks, arrangements of facilities, ways of transporting the heavy sacks, means of combining operations, and other possible improvements. This is exactly what *not* to do —immediately becoming entangled in details (of solutions, in fact) and not with the problem. In so doing you would be generating solutions to a problem you have never defined.

The current solution to a problem is not the problem itself. This may seem too obvious to say yet, if their behavior is an indication, many people apparently don't see the distinction. They attack the present solution, not the problem. There is a crucial difference between picking at a current solution in an effort to eliminate inadequacies and starting with a definition of the problem and from there synthesizing a superior solution through an intelligent problem-solving procedure. In the long run the latter will make you a much better engineer.

So now you know how *not* to go about it. How *should* it be done?

Problem Formulation

At the start, identify the desired transformation in terms of states A and B. This first step in the problem-solving process is *problem formulation*.

Problem formulation should be broad as well as first. To illustrate a broad problem formulation and why it is important, consider these alternative views of the feed problem:

To find the most economical method of . . .

Do you see differences in the solutions that might result from these various formulations?

1. **Transferring feed from *mixing bin* (state A) to *stockpiled sacks in the warehouse* (state B).**
2. **Transferring feed from mixing bin to *sacks on the delivery truck*.**
3. **Transferring feed from mixing bin to *delivery truck*.**
4. **Transferring feed from mixing bin to *delivery medium*.**
5. **Transferring feed from mixing bin to the *consumers' storage bins*.**
6. **Transferring feed from *storage bins of the feed ingredients* to the consumers' storage bins.**
7. **Transferring feed from *producer* to *consumer*.**

These formulations of the problem are not equally advisable; the probable consequences of pursuing each are quite different. In formulations 1 and 2 state B is *feed in sacks*. In transformation 3 only "truck" is specified, which opens the problem to solutions not involving sacks. In formulation 4 only "delivery medium" is assumed for state B, opening up additional possibilities. This trend toward a less specific definition of states A and B continues until only producer and consumer are specified, leaving the way open for a wide variety of methods of handling, modes of transportation, package types, and so on. From all this it is apparent that as the specifications assumed for states A and B become more general, the alternative solutions become more numerous and varied. Narrow problem formulations cause whole realms of profitable possibilities to be unnecessarily excluded from consideration. Most persons trying to solve the feed problem would automatically and unjustifiably assume state B to be stockpiled sacks in the warehouse; they would solve the problem without realizing that they themselves have limited the problem to this extent.

In formulation 1, state B goes only as far as the warehouse stockpile (Figure 4-3). Formulation 2 extends state B to the truck, and formulation 5 extends it to the consumer. In formulations 6 and 7 state A is extended. In these instances the problem is extended to embrace more of the total problem. In general, strive to formulate each problem *to include as much of the total problem as the importance of the situation and organizational boundaries will permit*. This is for a good reason. The more a total problem is split into subproblems to be solved separately, the less effective

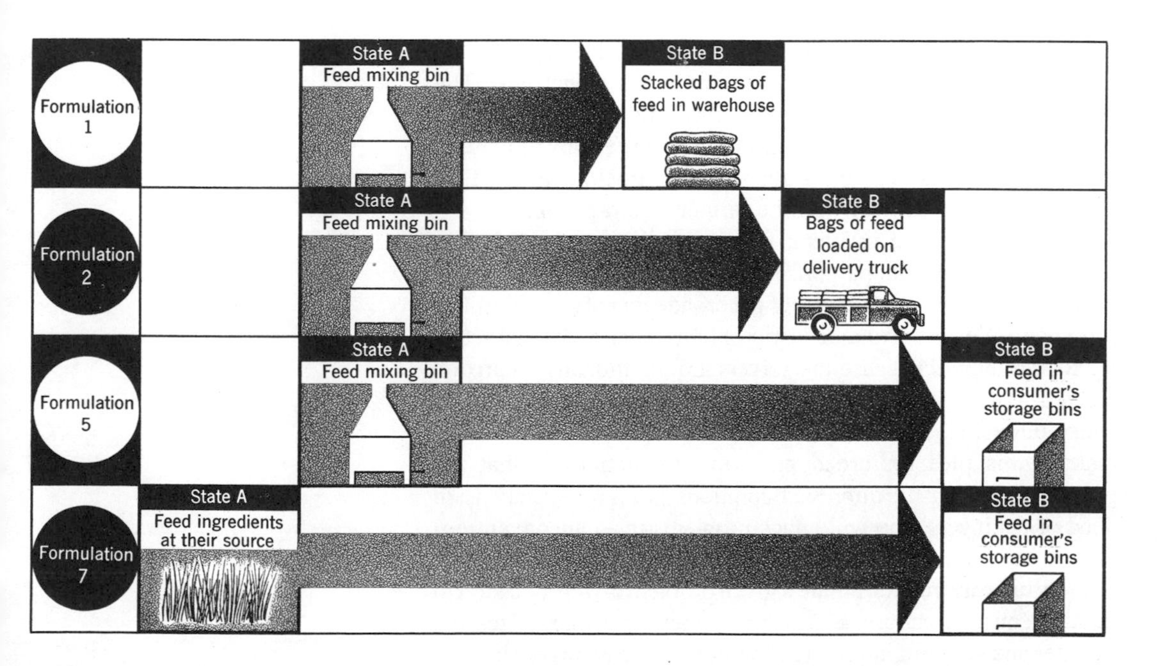

Figure 4-3 *Alternative formulations of the feed distribution problem, illustrating progressively broader formulations of a problem.*

the total solution is likely to be. If bagging the feed is treated as one problem, transportation to and stacking in the warehouse as another, transportation to the consumer as another, and unloading of trucks as still another, the overall feed distribution system that eventually evolves will probably be far from optimum. Treating this problem broadly is very likely to result in a superior solution overall.

The detail in which states A and B are specified and the proportion of the total problem they encompass will henceforth be referred to as the *breadth* of the problem formulation. Formulation 7 of this problem is certainly broad in contrast to formulation 1.

Importance of a Broad Formulation

The engineer assigned to this project succeeded in freeing himself from the limitation of using sacks and, thereby, he opened the problem to the possibility of handling feed in bulk. (Now you may think that's no feat. If so, you grossly underestimate the thought-restricting role of customary—especially long-standing—solutions to problems.) Also, his formulation encompassed delivery to the consumer, which opened the way to delivering feed in bulk directly into the farmer's storage bins. The result is that after many years of back-breaking loading and unloading of heavy sacks, dealers now deliver feed by blowing it through a tube leading from a bulk-delivery truck, a large "feed bin on wheels," directly into the farmer's storage bins.

Broad treatment of problems that previously were attacked in piecemeal fashion can pay off handsomely. Figure 4-4 provides an indication of what can come of the broadening of some familiar problems. We are surrounded by problems that are unsatisfactorily solved mainly because the solvers are traditionally "narrow-sighted." This applies to education, business, medicine, and every other field, as well as to engineering. The reason I am making this plea for broad problem formulations is that the probability of vastly improved solutions is high. Society is in need of engineers who will attack problems in an unconventionally broad manner.

How broadly you formulate a given problem is *your* decision to make. You can choose a very broad view that maximizes the number and scope of alternative solutions that can be considered. Or you can go to the other extreme and choose a formulation offering very little latitude in the way of solutions. Remember,

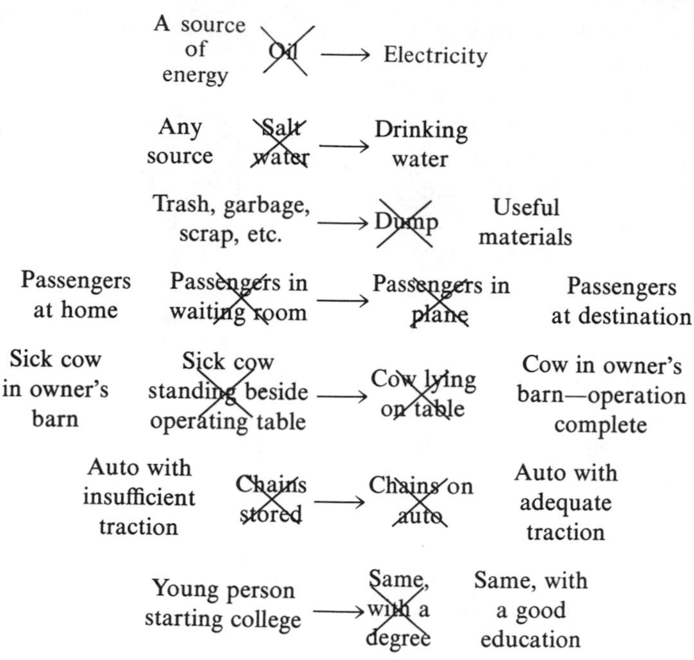

Figure 4-4. *Some formulations of familiar problems, indicating how each might be broadened. Perhaps you can visualize how these broadenings can pay off.*

however, a problem formulation is a point of view—the manner in which you perceive a problem. It may be no more than some thoughts or some scribbled notes. It is not irrevocable; it can be changed if you later find it necessary or desirable. Therefore you should formulate problems broadly; it is your prerogative—in fact, your professional obligation—to do so. You are selling yourself and your employer short if you don't.

Methods of Formulating a Problem

A problem can be formulated verbally or diagrammatically, on paper or in your mind. In many instances a few words will suffice. Or maybe you prefer a simple diagram. The black-box method of viewing a problem is a diagrammatic formulation. The usefulness of this approach can be illustrated by applying it to a type of problem that is often unsatisfactorily defined: an information-processing problem. An office that handles airplane reservations is an information-processing system. A customer comes to a ticket agent with a request in terms of number of seats, date, and perhaps other specifications. These constitute the input to the black box—state A, in other words. The output—state B—is also information, in the form of confirmation of the request or a quotation of the available alternatives. In the problem-formulation stage, what transpires within the black box is not known or of interest. The box replaces the details you are trying to avoid at this stage, and that is the key to its helpfulness.

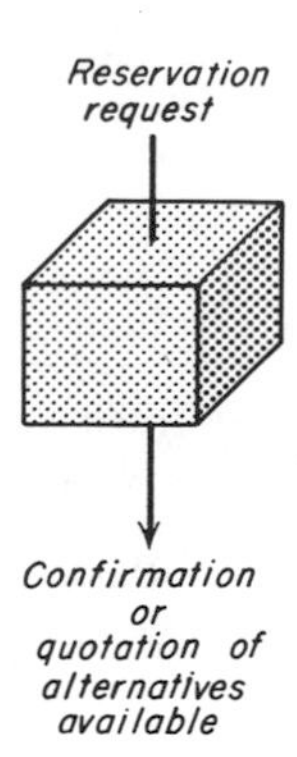

About now it must occur to you that there are no hard and fast rules for problem formulation. There is no such thing as *the* formulation of a given problem, but there certainly are more and less profitable ones. The best that can be done is to offer you guidelines and examples (Figure 4-4 should help in this respect). It is up to you to benefit from these and from your own experience in developing your problem-formulating skills.

From now on, when you are approaching a problem, hopefully you will, at the outset, carefully choose the problem you intend to solve and, in so doing, give thoughtful consideration to alternative formulations (Figure 4-3) *and* the probable consequences —payoffs as well as obstacles—of each. Strive for a broad, uncluttered view of a problem before you concern yourself with possible solutions, in fact, before you become entangled in *any* details. After you have done so the problem can be—must be—defined in detail.

Problem Analysis

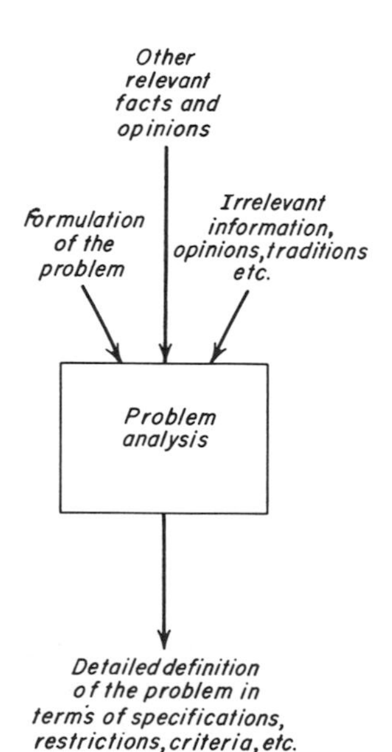

The following case illustrates defining a problem in detail. An appliance manufacturer has tentatively decided to market a new type of clothes washer. This machine is supposed to perform the usual tasks expected of such machines and also to serve as a home dry-cleaning unit. Management has also decided that:

1. **This unit *must not* be larger than 80 centimeters wide, 80 centimeters deep, or 100 centimeters high.**
2. **It *must* operate on 60-cycle, 115-volt alternating current.**
3. **It *must* be approved by Underwriters Laboratories.**
4. **The cost of manufacture *must not* exceed $125.**
5. **It *must* satisfactorily process all natural and synthetic textiles.**
6. **It *must* be foolproof against blunders in operation.**

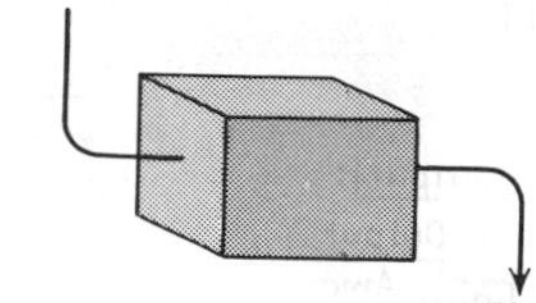

In formulating this problem, the engineer assigned to design this multipurpose cleaning machine identified state A simply as dirty fabrics and state B as clean fabrics. However, to solve this problem, he had to learn a lot more about it. He did, through considerable deliberation, investigation, and consultation, especially with marketing experts, who are in close touch with consumer preferences. The result of these efforts is the problem analysis shown in Figure 4-5. Here is an interpretation of his notes.

INPUT. (This engineer, as many others do, refers to what I have been calling state A as the input. Henceforth I will use state A and input interchangeably; likewise for state B and output.) An important part of his problem analysis is getting detailed information on the input and output. Here is an example of the detail required. Input and output tend to vary over time, which is certainly true for a washer. The amount of clothing a person places in the machine varies from load to load, and so do the types of fabrics and amount of dirt. This problem is typical in this respect; very few input and output characteristics remain constant in any problem. (Examples from other problems: the ore input to a steel mill varies in chemical makeup; the electrical output of a power plant certainly varies over a period of time.) These dynamic characteristics of states A and B are called *input variables* and *output variables*.

Often there is a limit on the extent to which an input or an output can fluctuate. In this problem an upper limit of 7 kilograms has been set for the input variable "weight of load." This is called an *input constraint;* the equivalent for state B is an *output constraint*. Before an engineer can solve a problem satisfactorily, he must have reliable estimates of input and output variables and their constraints.

RESTRICTIONS. A restriction is a solution characteristic previously fixed by decision, nature, law, or any other source the problem solver must honor. Nature dictates that light, water, and nutrients must be provided to transform seeds into plants. Through building codes, certain characteristics of structures are fixed by law. In the problem at hand, the decisions cited on page 70 are restrictions and appear as such in Figure 4-5. Each restricts the choices open to the problem solver. Some restrictions limit his choice to a *range* of values: for example, "the machine cannot be

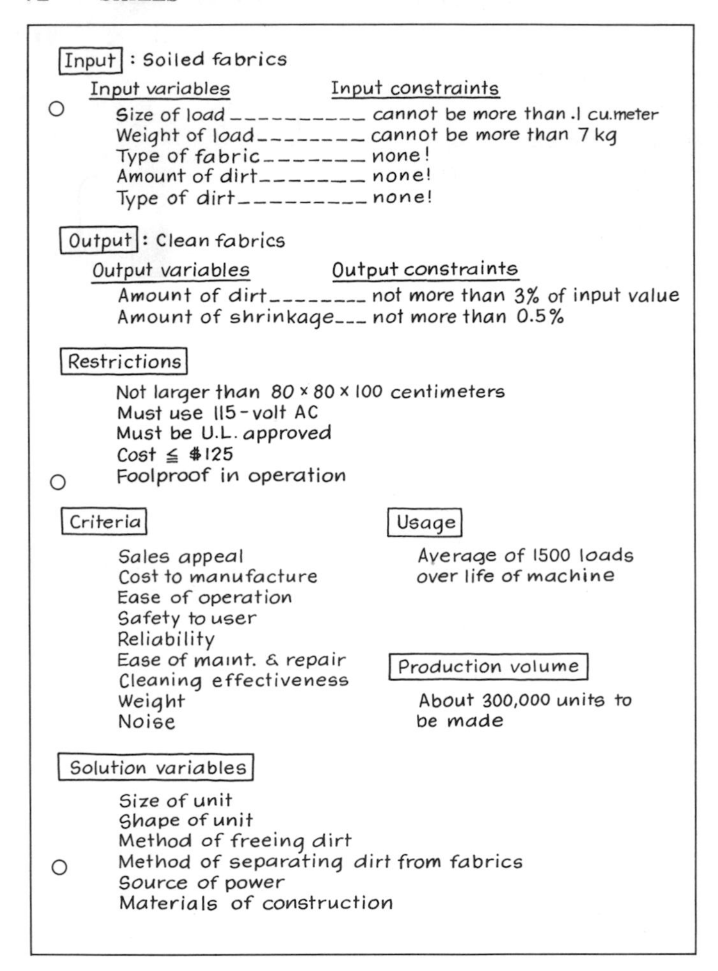

Figure 4-5 *This is a page from the engineer's notebook, summarizing his analysis of the cleaning machine problem.*

greater than 80 × 80 × 100 centimeters.'' Others *fix* a solution characteristic: ''it must operate on 60 cycle, 115 volt A.C.'' Therefore solutions larger than 80 × 80 × 100 centimeters are ineligible; so are solutions using other than the specified power source.

CRITERIA. The criteria to be used in selecting the best solution should be identified during problem analysis. Actually, criteria do not change radically from one engineering problem to another; construction cost, safety, reliability, appearance, ease of use, and maintenance cost are criteria that apply in almost every case. But the relative weights of these criteria are likely to change significantly. Hence, in most cases, the engineer's main task is to learn the relative importance attached to various criteria by officials, customers, clients, citizens, and others. This information is important, as the following case will illustrate. Assume that safety is to be a heavily weighted criterion in the design of a rotary lawn mower. Knowing this, the designer will consider a different set of materials, mechanisms, cutter types, and discharge methods than he would if safety were not considered extremely important. An unusually weighted criterion, therefore, affects the types of solutions the engineer will emphasize in the next phase of the design process, the search for alternative solutions.

Some readers may find the distinction between restrictions and criteria fuzzy. But note that a *restriction* is an either-or proposition; a solution either satisfies a restriction or it doesn't. With a *criterion,* being a value by which solutions are judged, it is a matter of degree; some solutions are more or less attractive than others. A criterion lacks the clear-cut, go/no-go feature of a restriction. We look for the *least* costly, the *most* attractive, and the *safest* when cost, attractiveness, and safety are the criteria. In the design of an appliance, to say that energy consumption matters is to say energy consumption is a criterion. Other things being equal, the solution that consumes the least energy will be chosen. To say that any solution that consumes more than 10 watts is unacceptable sets up a restriction.

SOLUTION VARIABLES. Alternative solutions to a problem differ in many respects. Solutions to the cleaning machine problem differ in size, shape, method of freeing dirt from fabrics, type of mechanism, materials of construction, and so forth. The ways in which solutions to a problem can differ are called *solution variables.* Strictly speaking, a solution variable is a solution characteristic that the problem solver is free to alter.

USAGE. Still another type of information that the engineer must obtain or predict himself is the extent to which his solution will be employed. This is often referred to as *usage,* and it very definitely affects the type of solution that will prove to be best in the situation at hand. Usage becomes significant whenever *total* cost—the cost of arriving at a solution *plus* the cost of physically creating it *plus* the cost associated with using it—is a matter of concern; and when is this not the case? If a river is to be crossed only rarely at a given spot, a bridge is obviously not the solution that minimizes *total* cost. On the other hand, if over a period of time millions of persons will cross the river at this spot, a rowboat is hardly the preferred method with respect to the total cost criterion.

PRODUCTION VOLUME. Suppose that only 10 of these cleaning machines are to be built. Under these circumstances the designer would care little about the manufacturability of his creation. If nonstandard, expensive components were specified and hand methods of fabrication were required, this would be of little concern as long as the volume was only 10 machines. However, 300,000 machines is a different story; under these circumstances the engineer will be vitally concerned with the manner in which alternative designs affect manufacturing cost. This number, referred to as *production volume,* has a significant effect on the type of solution that is best, and obviously the engineer should know what volume is expected before he starts thinking about solutions.

Problem Definition—Some Recommendations

Don't get caught solving a problem you have failed to define! Furthermore, define a problem in two stages, beginning with a broad, detail-free *problem formulation* and following it with a *problem analysis* in which your general statement of what is wanted is translated into specific problem characteristics. This two-stage definition counteracts the natural tendency to become immediately enmeshed in details, after which a broad perspective is virtually impossible to achieve.

Problem formulation is a crucial stage in an engineering project. It is here that the thrust of the project is determined. At this point you are deciding what problem you are going to solve. What more important decision can there be in the entire problem-solving process? What belies its importance is the fact that it may take only a matter of minutes.

Problem analysis, in contrast, is relatively time consuming, since it ordinarily involves a great deal of consultation, observation, thought, fact gathering, and negotiation over restrictions.

Incidentally, the treatment of problem analysis in this chapter is strictly introductory. Furthermore, the definition of terms is loose. Readers who are dissatisfied in either respect are referred to the appendix of this chapter.

The Process of Design

What you have been introduced to in this chapter are the first two phases of a five-phase problem-solving procedure that engineers generally refer to as the *design process* or simply *design*. In sum it consists of:

Problem formulation. **Broad, detail-free definition of the problem at hand.**

Problem analysis. **Detailed definition of the problem.**

Search for solutions. **Accumulation of alternative solutions through invention, inquiry, and the like.**

Decision. **Evaluation, comparison, and screening of the alternatives until the best one evolves.**

Specification. **Complete documentation of the chosen solution.**

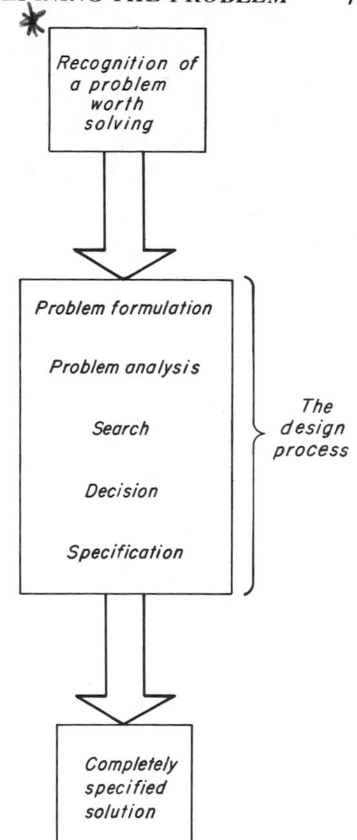

This discourse on problem solving continues in Chapter 5 by focusing on "the search."

MORE ON PROBLEM ANALYSIS

Appendix

The solution variable is the focal point of the engineer's problem-solving efforts. The challenge is to vary these solution variables in order to maximize the overall criterion. So, given:

- **a *criterion* (some say "measure of effectiveness"), which is a dependent variable, a function of one or more solution variables, and**

- ***solution variables,* which are independent (controllable, alterable) variables,**

it is the engineer's task to find the combination of values for the solution variables (S_i) that maximizes the criterion C.

Strictly speaking, a *restriction* is a limit on the range of values that can be assigned to a solution variable, established by prior decision, by law, or by nature. Three examples are:

$S_1 = f$, which fixes solution variable 1 at a quantity f
$S_2 < g$, which limits solution variable 2 to anything less than g
$h < S_3 < j$ which restricts solution variable 3 within h and j.

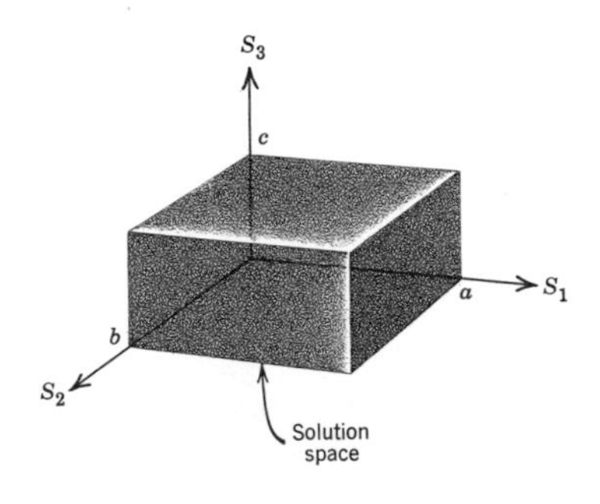

Figure 4-6

Picture a problem with three solution variables: S_1, S_2, and S_3. Suppose these solution variables are restricted so that S_1 cannot be greater than a, S_2 cannot be greater than b, and S_3 cannot be greater than c. These limits establish the *solution space,* shown in Figure 4-6. The final solution must come from this space. In most problems there are more than three solution variables, so the situation is one of N variables in N-dimensional space. When restrictions exist, there are boundaries on this multidimensional space equivalent to those in Figure 4-6.

Recall that in the cleaning-machine problem the cost of the unit could not exceed \$125. For simplicity this was listed as a restriction (and, in practice, I would be content to let it pass as that), but it is really a limit on a criterion, not on a solution variable. Any solution for which $C > \$125$ is ruled out. Such a limit on a criterion I call a *criterion constraint*. Thus, a restriction limits a solution variable; a criterion constraint limits a criterion.

Therefore, an engineer works with four types of variables in solving a problem. These variables, with their respective limits, are:

VARIABLE	*LIMIT*
Input variable	Input constraint
Output variable	Output constraint
Criterion	Criterion constraint
Solution variable	Restriction

How all this fits together is indicated in Figure 4-7, which summarizes the types of information needed for a complete problem analysis. Given this information, it remains for the engineer to find the combination of values for $S_1, S_2, S_3, \ldots, S_N$ that maximizes (or minimizes, if such is appropriate) C *and* satisfies all restrictions and constraints. To put it more rigorously, the engineer seeks to optimize

$$C = f(S_1, S_2, S_3, \ldots, S_N)$$

subject to all restrictions and constraints.

Input, with input variables
$I_1, I_2, I_3, \ldots, I_N$
and these constraints:
$I_1 = a$
$0 < I_2 < b$
$c \leqq I_4 \leqq d$
etc.

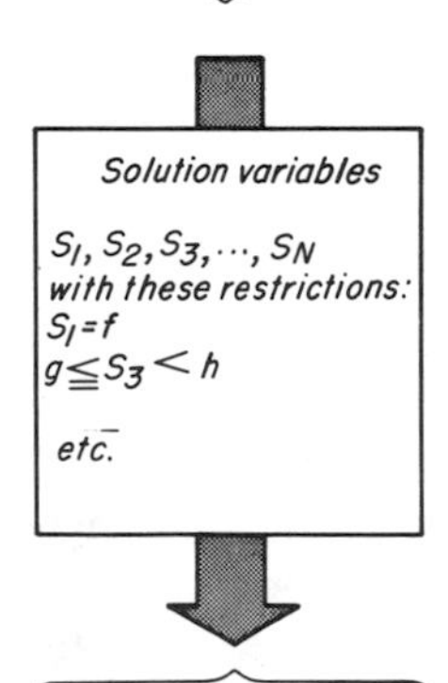

Output, with output variables
$O_1, O_2, O_3, \ldots, O_N$
and these constraints:
$O_2 > j$
$O_5 < k$
etc.

Criterion = C
Usage = U
Production volume = V

Figure 4-7 *Summary of the types of information you gather in a problem analysis. Lowercase letters represent what would ordinarily be numbers.*

EXERCISES

1. Complete the following formulations.

Businessman's problem:	An invested sum of money	⟶	The (eventual) return of that sum plus a profit
Missionary's problem:	A tribe of cannibals	⟶	
Defense lawyer's problem:	Twelve undecided persons	⟶	
Urban planner's problem:	A physically, socially, and economically blighted urban area	⟶	
Newspaper boy's problem:		⟶	
Mountain climber's problem:		⟶	
TV repairman's problem:		⟶	

2. Assume that you are designing the following:
   (a) Vegetable cannery.
   (b) Structure for crossing the English Channel.
   (c) Transoceanic telephone cable.
   (d) Power-generating station using coal as fuel.

How do you formulate each of these problems? What information concerning input and output would you gather during your analysis of each problem? For each, list some major criteria that you believe should be employed.

3. Identify states A and B for the prime problem faced by each of the following: cook, firefighter, teacher, pottery maker. Make assumptions where necessary.

4. Formulate the problem to which each of the following is a solution: gearbox, electric iron, air conditioner, oil pipeline, bottling plant, snowplow, commuter train, public address sys-

tem. (For example, the usual input to a gearbox is a shaft rotating at one speed and the output is a shaft rotating at a different speed.)

5. *Formulate* the harvesting problem assigned to Stu Dent via the memo of Figure 4-8. In fact, cite alternative formulations (three seems a reasonable number) and then discuss the relative merits and probable outcomes of each formulation. Your formulations can be in words, diagrams, or sketches.

6. *Analyze* the apple-harvesting problem characterized by Figure 4-8 on the basis of your preferred formulation. I suggest that you pattern your analysis after Figure 4-5.

7. *Define* the problem introduced by Figure 4-9. You are expected to defend your formulation. Consult your instructor if you want additional information.

8. Cartons moving on a belt conveyor are spaced irregularly. As they approach the labeling machine, they must be spaced at regular intervals. Formulate this problem. Analyze it, based on information supplied by your instructor.

9. At a clinic for large animals, a number of operations must be performed on horses, cows, and bulls. As you can well imagine, getting one of these animals to lie on an operating table is a challenge. The head administrator of the clinic has approached your firm with a request that has been turned over to you. He wants hoist-and-sling arrangements installed to lift animals onto their operating tables, and it is your assignment to design them and their mountings. Formulate and then analyze this problem. Include a brief paragraph defending the formulation you recommend.

10. Define the problem characterized by one of the case histories in Chapter 1.

DESIGNWELL ASSOCIATES

Consulting Engineers Inter Office Memo

TO: Stu Dent, Project Engineer

FROM: James C. Hogan, Chief Engineer

The following request for a design proposal was received from Juicy Fruit Orchards, one of the largest growers of apples. I quote:

> "We are seeking significant improvements in our apple harvesting methods. A severe labor shortage and rising labor costs have prompted us to seek proposals for harvesting systems that require less labor. To be competitive with current hand-picking methods, the harvesting cost of an alternative system cannot be greater than $.20 per bushel."
>
> "Juicy Fruits harvests approximately 900,000 bushels of apples that are converted to a variety of apple products at a nearby processing plant."
>
> "Most of our apple trees are between 4.5 and 6.0 meters tall, and an average of 6.3 meters in diameter. Apples are removed from the orchard area in large pallet boxes, 112 x 112 x 81 (height) centimeters. Not more than five percent of the apples sold whole for eating purposes may be bruised. Not more than thirty percent of the processed apples may be bruised. When harvested, the tree is stripped in one picking. The bulk of the harvesting is done between mid-September and mid-November."

You are assigned to this project. I expect a preliminary report in the very near future.

Figure 4-8

DESIGNWELL ASSOCIATES

Consulting Engineers Inter Office Memo

TO: Stu Dent, Project Engineer

FROM: James C. Hogan, Chief Engineer

A client has engaged us to prepare a design concept for a solid-waste sorting system that will receive residential waste from standard collection vehicles and sort it into glass, iron and steel, aluminum, paper, and residue (mainly plastics and decomposible material). All outputs will go into Dempster Dumpster bins except paper, which is to be baled. The client insists that this facility operate with a reliability of .95 or better.

I am assigning you to the design-concept phase of this project, with the expectation that you will have a report for me in the very near future.

Figure 4-9

DESIGN: THE SEARCH **Chapter 5**

A vast store of technical information does not, by itself, make you valuable to employers and society. The payoff comes when you apply this knowledge—solve problems—or more specifically, when you *create* structures, machines, devices, or processes that satisfy human needs and wants. So, since engineering is basically a creative activity, a person who is prepared for it has read and thought about creativity and invention and has worked at developing his inventive skills. You can begin such efforts right now by carefully considering the contents of this chapter. It is intended to focus your attention on the nature and importance of the creative act in engineering, to inspire you to enhance and capitalize on your creative abilities, and to show you how to do both.

Solutions occur to you quite unpredictably when you are laboring over a problem. You may get an idea the moment you learn about the problem, or during Sunday's sermon, or on your way to work, or three days after you've already "solved" it, or. . . . This is the way of the mind; why fight it? But subconscious, leave-it-to-chance methods can hardly be relied on to solve consequential problems satisfactorily and consistently. What is called for is a deliberate and energetic search for alternative solutions in which conception of solutions is the objective rather than a by-product of your efforts. Therefore, after you have defined a problem, you must turn your efforts to what is truly a search—of the mind, the literature, and the world about you—for solutions.

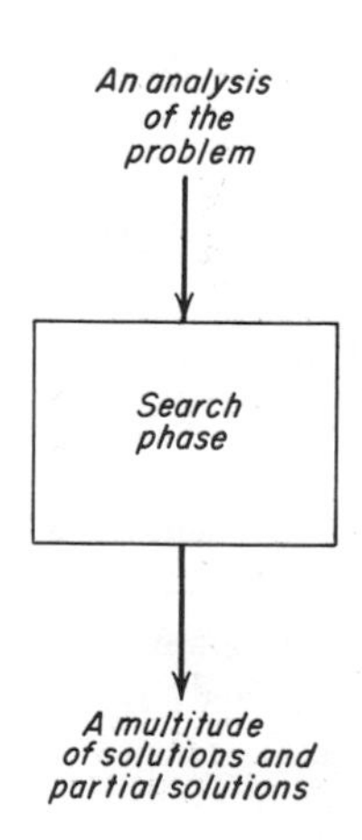

Surely you have wondered where your solutions will come from when you are a practicing engineer. Probably the first, maybe the only, but certainly a logical source to occur to you is "books." True, man's vast accumulation of knowledge provides ready-made solutions for some parts of most problems. Searching for these is a relatively straightforward process of exploring textbooks, handbooks, technical reports, journals, existing practices and, of course, your own memory. But there is a second major source of solutions—your own ideas—the fruits of the mental process called *invention*. You *will* be relying heavily on your ingenuity to solve many aspects of problems not covered by existing technical and scientific know-how. Unfortunately, however, inventing solutions is not as straightforward and controllable as looking up ready-made ones; you may recognize this from your own problem-solving experience. So it pays to devote special attention to improving your inventive ability.

For some of you, learning that engineering work offers ample opportunity to exercise your ingenuity is a happy thought. It may make others apprehensive, but there is no need for this. Almost all of you have—or can develop—what it takes to meet the creative demands of your job. That, of course, is an opinion, but a well-founded one. Hear my argument.

Your inventiveness as an engineer will depend on: *inherited qualities;* your *attitude;* your *knowledge;* the *effort* you exert; and the *method* you employ in seeking out ideas. This leads to a significant point. *Since you control all but the first of these five determinants, it is within your power to improve your inventive ability.* You *can* over a period of time, develop a more favorable attitude and increase your knowledge. You *can* increase your effort. You *can* substantially improve your method of searching for solutions. This should be good news to you if you were not born a creative genius; it means you can compensate for this through your influence over the remaining four determinants. Some elaboration is in order.

The Influence of Attitude

You won't get far in your efforts to become more creative if you don't believe in yourself. To succeed, you have to be convinced that you can be as inventive as an engineer need be. Hopefully, you will be convinced by the following argument.

Why are some persons noticeably more inventive than others? "Born that way" is what you are probably thinking. Hogwash! On what basis do you conclude that heredity accounts for this when it is only one of the *five* determinants of a person's creative performance? An unusually creative person could just as well be that way not because he was exceptionally endowed at birth but because his attitude, knowledge, or method is exceptional, because he works hard, or for any combination of these reasons. It may well be that a rare few are exceptionally endowed through heredity, but it is also true that *everyone inherits inventive ability and that very few people fully exploit the potential they have*. It is very likely, as far as this trait is concerned, that you weren't shortchanged at birth. So don't be concerned about whether you have aptitude for invention; concern yourself with putting what you have to use.

It will also help if you can develop an insatiable drive to find better solutions. Rarely, if ever, will you work on an engineering project in which you are able to identify all solutions. Usually you will run out of time long before you run out of possibilities. The engineer who developed a baggage handling system for an air terminal conceived various methods of transporting passengers' luggage, numerous sorting systems, and many pick-up arrangements. He finally ended his search because other problems were awaiting his attention, not because he had exhausted all possibilities. Nor did he believe at any time during this project that he had thought of all solutions. He assumed that more and better ones remained, and he went after them. He is what we call "alternatives hungry."

If you are to be that way you must become alternatives conscious. One way of achieving this is to bombard you with problems to which there are impressive numbers of alternative solutions, starting with Figure 5-1. If this produces the desired alternatives consciousness, there is little chance you will be caught "solving" a problem without uncovering and considering a reasonable number of alternatives. Each time you isolate an additional solution, set out again to find a better one, continuing until you are forced to give up because of a project deadline or the pressures of other problems awaiting your attention.

The Role of Knowledge

What we loosely refer to as an idea is the combining of two or more bits of information in a manner that is novel to the conceiver. It is a reorganization of knowledge; no idea evolves from nothingness. And so the larger your store of knowledge is, the more "raw material" you have from which you can synthesize solutions. Furthermore, the broader the range of subjects this knowledge embraces, the better the prospects for unique and often particularly effective ideas. School is not the only source of such knowledge; observation, conversation, reading, and other forms of lifelong learning are important, too.

Figure 5-1 *Speaking of alternative solutions, here are* some *of the methods devised for powering a plane in vertical flight. These, plus the familiar helicopter, plus the VTOLS shown in Chapter 1, are being considered for military and commercial use.*

Propulsive wing

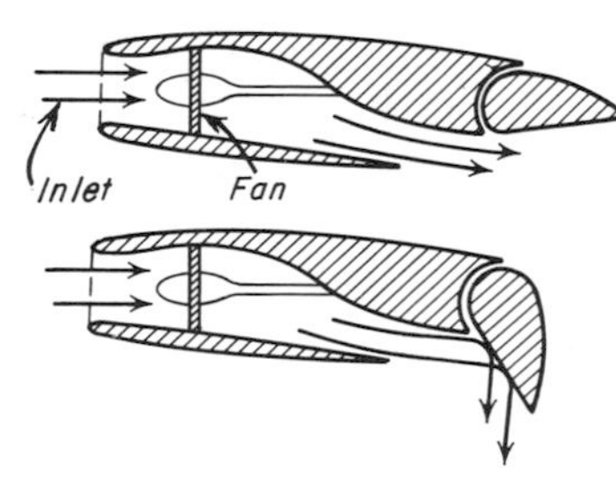

Deflected jet

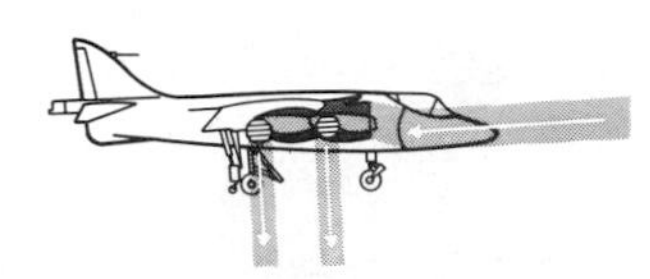

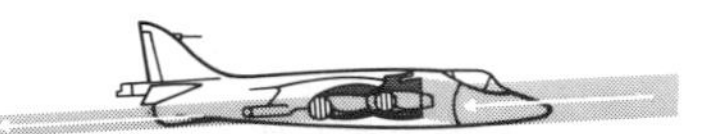

Wing tip engine pods

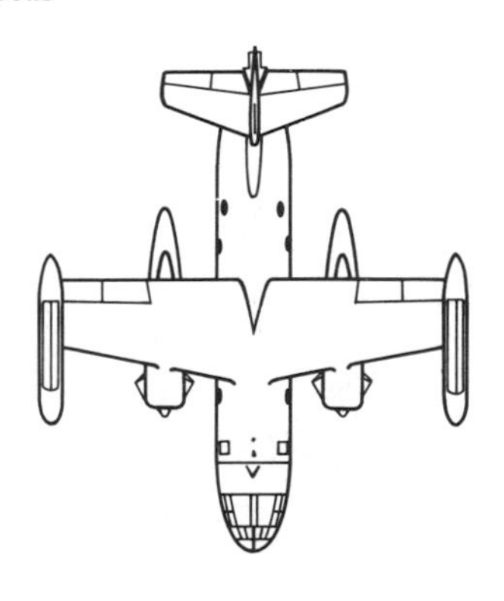

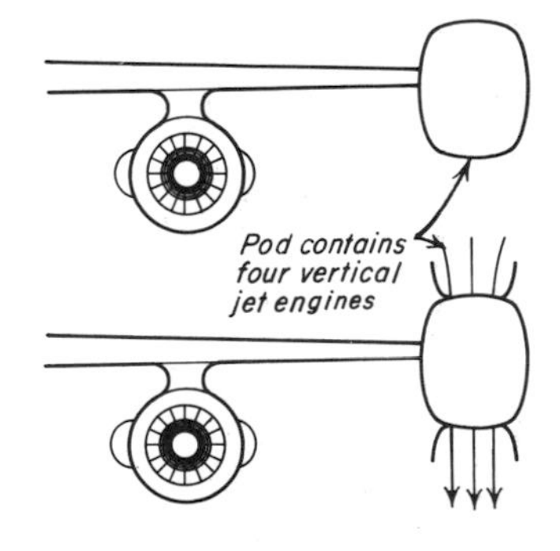

Vertical fuselage engines

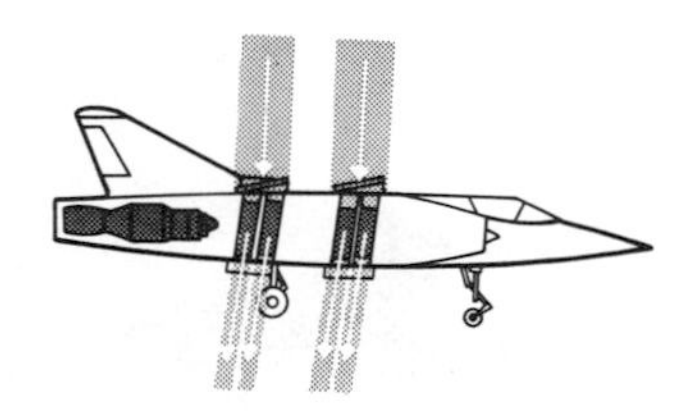

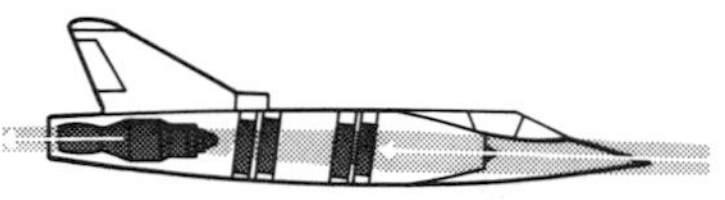

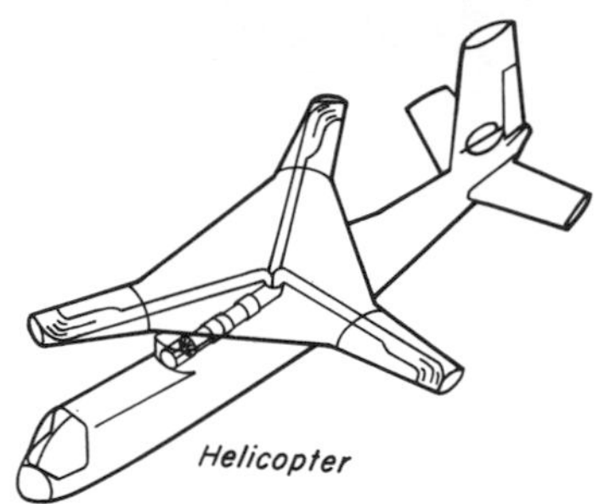

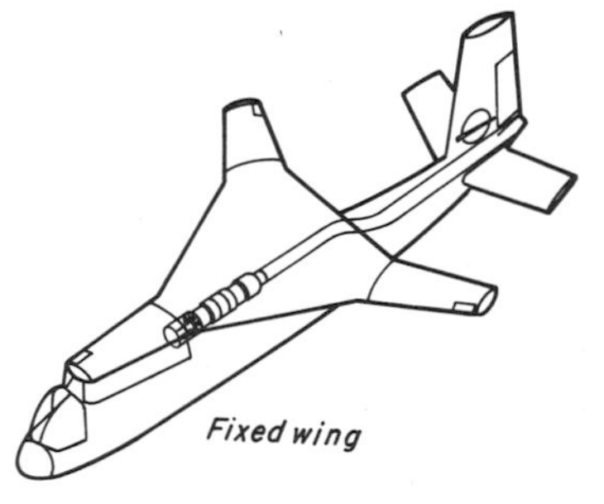

Rotating wing

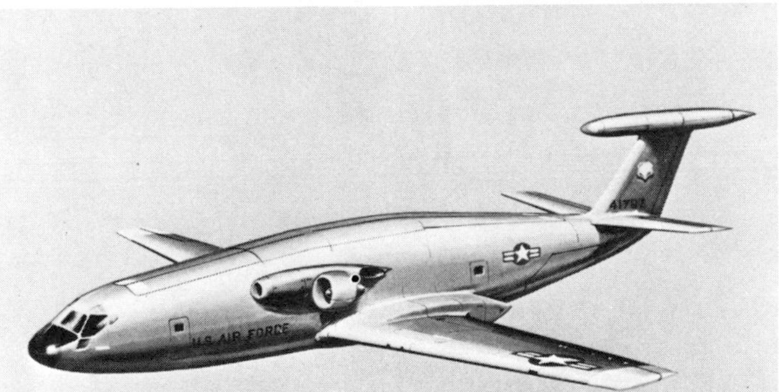

Stowed rotor

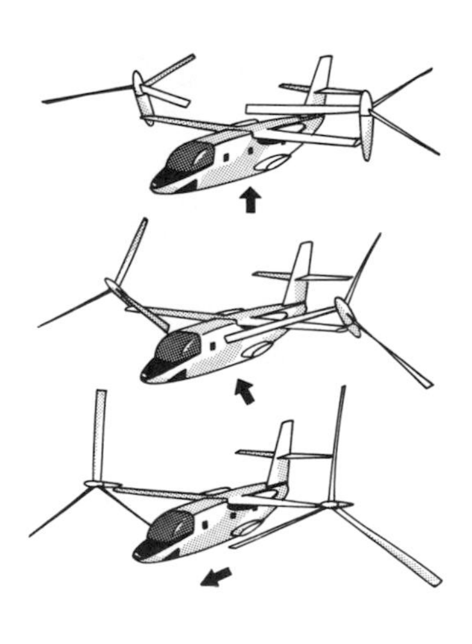

Tilting rotor

Ducted propeller

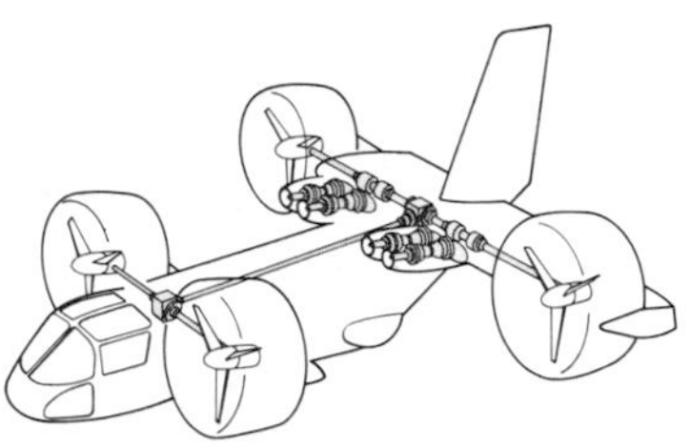

The Influence of Effort

There is a big difference between an occasional flash of genius and consistently producing, under pressure, ideas and more ideas, to solve a given problem in the limited time available. The latter takes effort. You won't find many really creative persons who are not hard workers. So, to maximize your inventiveness, you must be willing to think long and hard, even to sweat a little.

The Influence of Method

The following analogy highlights some of the pitfalls, difficulties, and procedural flaws you are vulnerable to when seeking solutions to a problem. Visualize the *X*s pictured in Figure 5-2 as points in space, each representing a solution to a

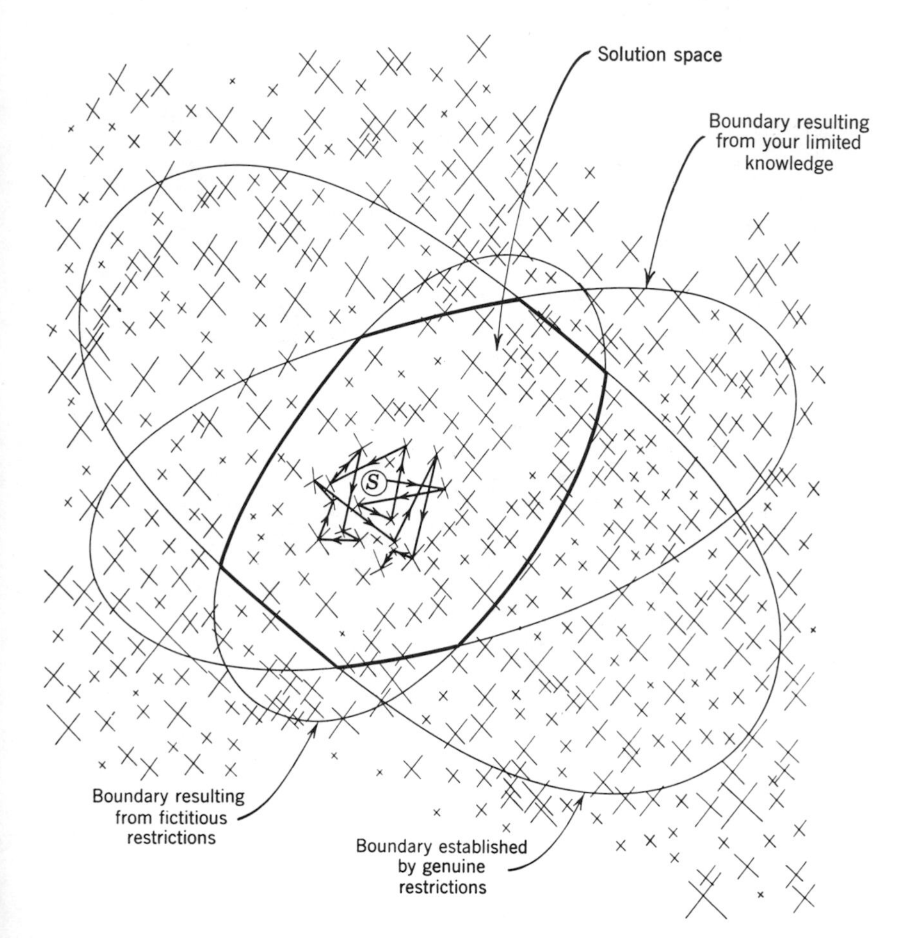

Figure 5-2 *The arrowed path represents the manner in which you are prone to search for new solutions to a problem unless you have devoted careful attention to your search method.*

problem at hand. Assume that widely separated points represent radically different solutions. When trying to think of solutions to a problem, wouldn't it be nice if you could start somewhere in this solution space and move to progressively better solutions until perfection or, more likely, a deadline ended your search? Unfortunately, that's not the way the mind ordinarily works. Instead, your search is likely to be hit or miss, suffering from objectionable degrees of regression, inefficiency, and lack of direction.

You may be thinking, "Not I." Yet you would probably be appalled if you carefully and objectively examined what you actually do when seeking ideas. All too often you start with the presently used solution (the point *S* in Figure 5-2) and proceed from one point (idea) to another in a manner indicated by the arrowed path. Notice that the jumps tend to be relatively small, so that ideas tend to cluster about the current solution.

Why do solutions tend to closely resemble the present way(s) of solving the problem? Mainly because the customary solution to a problem has an almost uncanny power of attraction. This power is especially strong if that solution has a long history of use. After you have "lived" with a certain solution for a while and have become so accustomed to it, your thinking is bound to be in a rut. History abounds with illustrations, an excellent one being the record of man's attempts to fly. The solution men were familiar with was the wing action of birds and insects. Almost inevitably, then, they tried to fly by employing all sorts of unsuccessful and often disastrous wing-flapping schemes. In time they freed their thinking from "the stranglehold of the familiar."

This analogy (Figure 5-2) will help you to appreciate why it pays to devote careful attention to your method of searching for solutions. It indicates a few of the tendencies that inhibit your ability to invent, *unless* you take measures to minimize their effects. Fortunately, such measures exist. Give them serious consideration; they are probably your best bet if you wish to improve your inventiveness. They are:

First, maximize the number and variety of solutions on which you can draw by enlarging the solution space as far as practical. **You do this by pushing back the boundaries that establish that space (Figure 5-2). There are opportunities to do so in almost every problem: by ridding yourself of imagined and self-imposed restrictions, by shaking off unjustified "real" restrictions, and by supplementing your knowledge relevant to the problem at hand.**

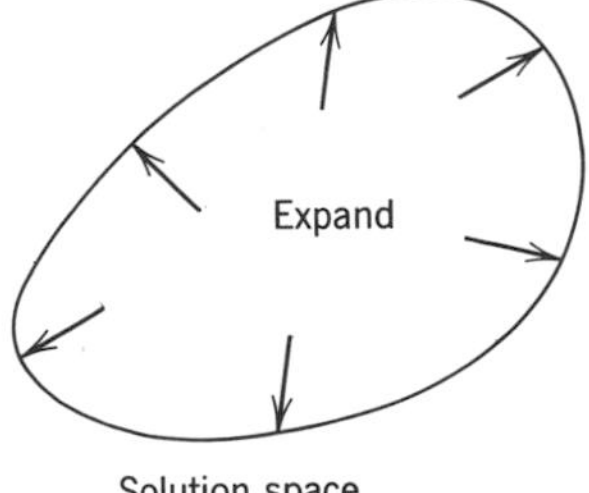

Then, take full advantage of this expanded solution space; search it effectively. **You should sample all areas of possibility that offer promise of containing the optimum solution. This requires that you triumph over the pitfalls and temptations cited with the aid of Figure 5-2. I can help you with a strategy found to be very effective: making your search partially systematic and partially random. Both approaches warrant elaboration.**

ADDING A SEMBLANCE OF SYSTEM TO YOUR SEARCH. An excellent way of doing so is illustrated here. An engineer is engaged in the design of an improved system for harvesting apples. In the course of defining the problem he identifies solution variables such as "method of separating apples from the tree," "method of bringing the separator (e.g., a man) to the apples," and "method of collecting the separated apples." As he concentrates on the solution variable "method of separation," he searches first for *basic* methods of separation; then he looks for specific versions of each of these basic alternatives, as illustrated by Figure 5-3. This minimizes the chance that he will overlook a whole block of promising possibilities. He then concentrates on other solution variables in turn, attempting to accumulate as many possibilities for each, always working from the general to the specific. It is doubtful that you will find a more effective method of improving your inventiveness.

The alternative solutions generated for a solution variable (e.g., those in Figure 5-3 for "method of separation") are referred to as *partial solutions*. Eventually the engineer will evaluate these partial solutions and combinations of them, perhaps recombining and reevaluating numerous times, until he has synthesized a complete solution that is the best combination of partial solutions.

There are other ways of introducing system into your search. You can systematically concentrate on each criterion, trying to generate, for example, means of maximizing reliability or of minimizing construction cost. You can systematically rearrange partial solutions in hopes of finding a high-payoff combination. The *alternatives tree* (Figure 5-3) is an effective means of systematizing your thinking. Any other way of organizing your thoughts and investigations so that a wide range of basically different solutions is brought under consideration is likely to be profitable.

RANDOM METHODS. There are some predominantly random methods of getting the mind into what might otherwise be "unexplored territory." Notable is the technique of *brainstorming*. A half-dozen or so persons assemble for the purpose of generating solutions to a problem. The leader describes the problem and the participants then bombard him with ideas, which he records on a blackboard. The objective is to accumulate many ideas by creating an atmosphere that encourages everyone to

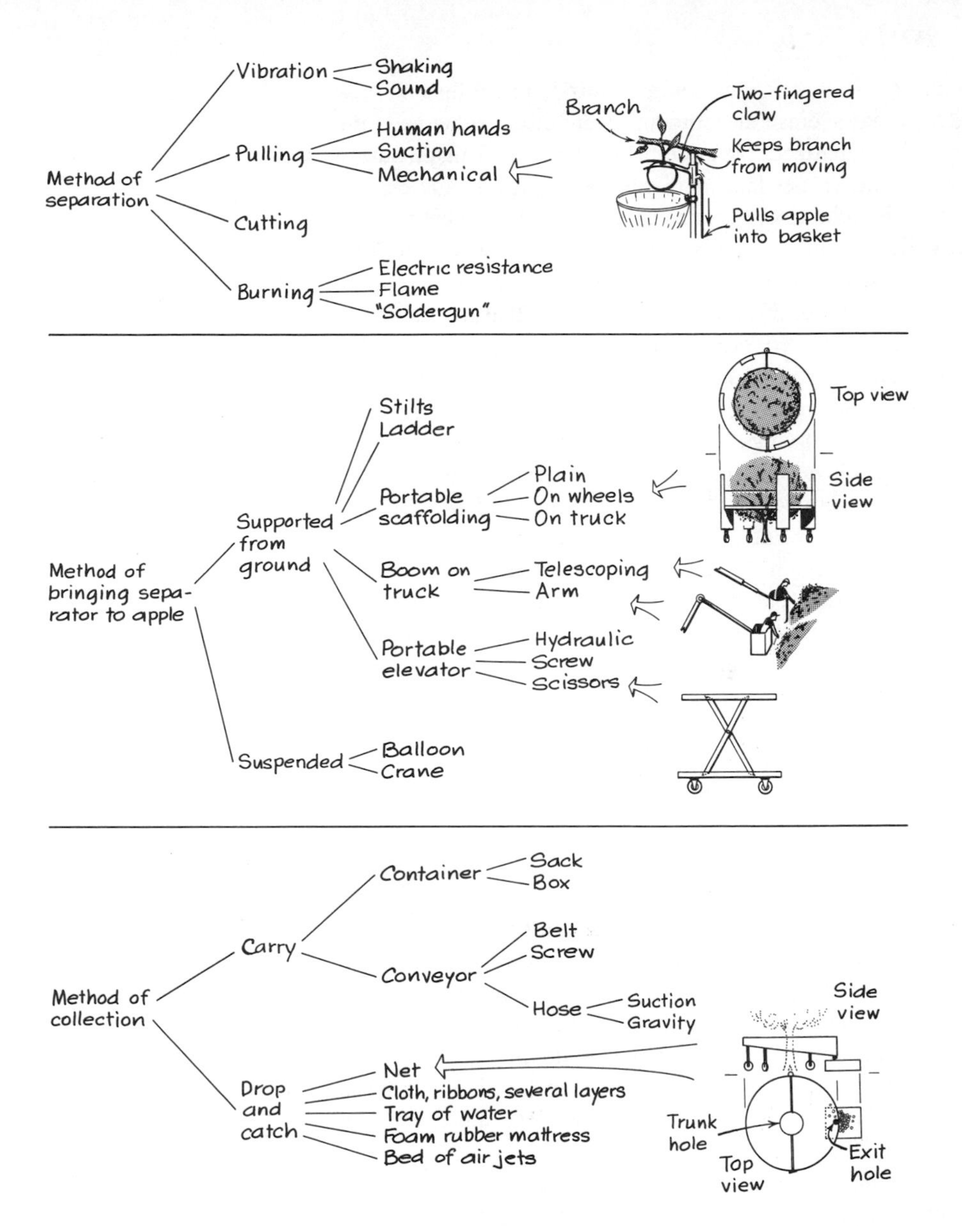

Figure 5-3 *A page from the engineer's notebook, showing some of the alternative partial solutions he generated for the apple-harvesting problem. For each major solution variable, he classified his ideas in the form of an alternatives tree.*

contribute all solutions that come to mind, regardless of how absurd they may seem at the moment. Ridicule and other negative responses are penalized. No attempt at evaluation is made during the session; this comes later, and wisely so. After some experience with this technique, a group can achieve a freewheeling flow of ideas that yields an impressive number and variety of solutions, often in 15 minutes or less.

One reason that brainstorming is fruitful is that the rapid-fire flow of ideas is repeatedly redirecting each participant's thoughts into new channels. In terms of the model pictured earlier (Figure 5-2), each person's mind is buffeted about the solution space in random fashion, forcing "big jumps" to distant points, thus combating the clustering tendency. The probability is high that some of these excursions will lead to profitable ideas.

Now you know what is meant by a "partially systematic, partially random search strategy." It is intended, one way or the other, to guide your search for solutions into profitable areas of possibility you might otherwise overlook. Since systematic methods have their limits and since you certainly do not want to rely only on a random search, it pays to call on both approaches.

Two Important Don'ts

Don't get bogged down with details sooner than necessary. Suppose you start working out the details of the first "good" idea you have. For all practical purposes your search will probably end right there. You will be spending time on details when you should be searching for other basically different solutions. Furthermore, preoccupation with the details of one solution severely hampers your ability to think of significantly different ones. Also, if you fall prey to this temptation and you happen to uncover a superior solution later, you will probably be unjustly biased in favor of the solution in which you have already invested so much time on details. And, finally, many alternatives can be satisfactorily evaluated while still in a relatively crude state of specification; since most will be rejected, why waste time on detailing them?

Therefore postpone details until they become necessary for decision-making purposes. In fact, it is best to form only solution concepts in this phase of the design process. (A *solution concept* is the essence, the gist, the general nature of a particular solution. Its form may be a rough sketch, a few words, or a sentence or two. The ideas shown in Figure 5-3 are partial solution concepts.)

Avoid premature evaluation; it has the same detrimental effects as premature preoccupation with details. This is the *search* phase of the design process; it is followed by the *decision* phase, in which evaluation of alternatives predominates. Therefore, ideas will not go unevaluated, but good ideas may go undiscovered if you become preoccupied with evaluation when you should be searching for better solutions.

Enough talk about techniques for improving your inventiveness. Perhaps you are convinced by now that there *are* things you can do about it. Certainly much more could be said about the process of invention, particularly its stimulation, and much more *has* been said. It is a popular topic for books, articles, and papers, so dearth of literature is certainly no obstacle if you wish to explore this subject further. There are suggested references in the Annotated Reading List on page 339.

A Plea for More Originality

These parting comments on the search phase concern what has become a cause for me: promotion of more imaginative thinking on the part of engineers. Too many solutions are the offspring of the past; too few are the products of original thought. Inertia perpetuates a host of inferior solutions in the world about us, leaving lots of running room for young engineers—hopefully, you will be among them—who will rely more on inventiveness.

There are explanations for this undue reliance on solutions from the past. It is certainly the path of least effort. Moreover, habitual solutions, although imperfect, are at least proven, whereas the performance of radical departures can be discomfortingly uncertain. Furthermore, you are likely to find that your problem-solving efforts become less imaginative as you become more knowledgeable in your field of specialization. It's a fact of life—the more you learn through education and experience about the accepted manner of doing things, the more constrained your thinking becomes. This is why some of the most imaginative thinking in a field often originates from people who are new, sometimes virtual "foreigners" to that discipline. How often I have assigned a simple engineering problem to freshmen engineers only to have them moan, "How can we solve a problem like this? We don't know anything about it." Little do they realize that their ignorance is in their favor as far as originality is concerned. They usually produce some imaginative solutions; they *can't* fall back on the customary.

It is easy to miss the point here. No one is knocking the vast store of valuable know-how that man has accumulated or in any way suggesting that you avoid it. The point is this: by all means acquire specialized knowledge, use handbooks, learn through experience how things are done. But draw on this store of customary solutions *and* on your capacity for original ones; both have much to contribute. The trick is to be creative in spite of your specialized knowledge, since without that you can't get along in modern engineering.

You can derive more than the obvious from this chapter's "creativity message." In particular, you should have gained further insight into the practice of engineering and preparation for it. Perhaps you were not fully aware of the need for creativeness in engineering and the attention given to it by engineers. Many persons are not, which reminds me of a frustrating experience. For years I have been promoting the idea of a campus seminar on the subject of creativity. In it architects, artists, musicians, poets, authors, scientists, mathematicians, engineers, and others would exchange views of creativity, its nature, enhancement, and perception. All are artists to an extent, and surely an exchange on this subject would be a stimulating experience. But enough of this vision; the point of mentioning it is the reaction I get from my nontechnical colleagues. They are obviously surprised, and can't quite get why an engineer is interested in creativity. Nor can they envision what engineers might contribute to such a seminar. Such responses reveal one of the common false impressions of engineering.

EXERCISES

1. For one of the following problems, identify a major solution variable, then generate as many partial solutions as you can for that variable.
   (a) Snow removal from city streets.
   (b) The tree transplanting problem.
   (c) Coin sorter-counter.
   (d) A problem specified by your instructor.

I urge you to follow the pattern established by Figure 5-3.

2. With between five and ten students, generate as many ideas as you can during a fifteen-minute brainstorming session. Some potential subjects: ways of getting more ideas; uses of flywheels;

means of removing an oil slick from a body of water; or a topic your instructor suggests. One or two members of the group should record suggestions on the blackboard. Remember, no evaluation, or at least no signs of it. (Without the enthusiastic participation of most members of the group, this experience is going to be a bomb.)

3. Go to work on the ''animal lift'' problem (exercise 9, page 79). See how many alternative solutions you can conceive. Details are not expected; words and sketches will suffice. Surely you will generate a number of partial solutions for each of the major solution variables. You may wish to synthesize a proposed system from the best of these partial solutions.

4. Do as instructed in exercise 3 for the trash sorting problem (Figure 4-9).

5. Generate alternative solutions for the iceberg problem (Figure 3-2c). How do you move an iceberg several hundred or more kilometers to water-needy areas?

6. In spite of signs and more signs, drivers still insist on driving onto high-speed, dual highways via an exit. You can imagine the usual consequences. How many schemes can you come up with that reduce the probability of this happening, or better yet, eliminate the possibility (i.e., make the exits foolproof in this respect)?

7. How many alternatives can you identify for the carton-spacing probem? (Exercise 8, Chapter 4.)

DESIGN: THE DECISION PHASE

Chapter 6

In the search phase an engineer expands the number and variety of solution candidates, as shown in the upper half of Figure 6-1. What is needed next is an elimination procedure that will reduce these alternatives to the preferred solution, as pictured in this figure and described below.

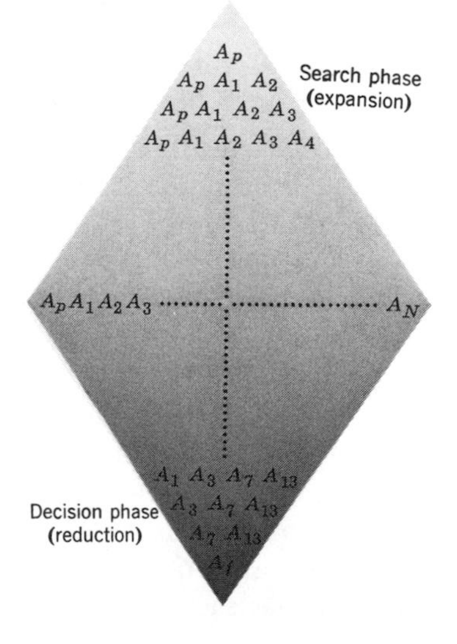

Figure 6-1 *A representation of the search and decision phases of the design process. The symbol A represents an alternative solution. The search phase begins with the present solution; the decision phase ends with the final solution.*

THE DECISION PHASE

An automobile manufacturer has for some time been using an expensive machine to mount tires on wheels. Because of an increase in the number of mountings to be made, an engineer was asked to do something to speed up the process. The result of his efforts is a new and much simpler device (Figure 6-2). Before proposing this device to the management, he made a thorough comparison of his solution and the existing one. The results are shown in Figure 6-3, which shows the criteria on which he based his decision, how the alternatives stack up with respect to these criteria, and the tabular summary he prepared to facilitate comparison. This case is unusual in that it involves only two major alternatives, but its uncomplicated nature makes it an excellent illustration for the following introduction to decision making in engineering.

The General Decision-Making Process

Although the specifics vary from situation to situation, in almost every instance the following steps must be taken before an intelligent engineering decision can be reached: (1) criteria must be selected and their relative weights determined; (2) performance of candidate solutions must be predicted with respect to these criteria; (3) the candidates must be compared on the basis of these predicted performances; and (4) a choice must be made. Each deserves some explanation.

1. Usually the overriding criterion is the *benefit-to-cost* ratio, which is the benefit expected from a solution relative to the cost of creating and using it. For the tire mounter the *benefit* is the saving in operating cost made possible by the proposed device. The *cost* is the total expense incurred in building and installing it (bottom of Figure 6-3). The benefit-cost ratio being employed in deciding on a proposed dam is

$$\frac{\textbf{Flood protection + conservation and recreation benefits + \ldots}}{\textbf{Cost of land + construction cost + maintenance cost + hardship to displaced persons + \ldots}}$$

The people who are footing the bill are interested in the expected tab as well as the prospective benefits of a proposed engineering venture. In fact, claims of benefits mean little if the associated costs are unknown. Knowing that a certain machine will cut his harvesting costs by X dollars a year does not mean much to a farmer until he also knows the cost of that machine. As a taxpayer, you insist on being told the cost of a proposed public works project before passing judgment on it. Similarly, an engineer rarely describes the benefits attributable to a proposal without also quoting the cost of bringing about those benefits.

A term used synonymously with benefit-cost ratio is *effectiveness-cost ratio*. Furthermore, some authors speak of cost-benefit instead of benefit-cost, and others talk of *return on the investment* but, regardless of the specific terms or their ordering, the notion is the same important one: total benefits relative to and inseparable from total cost.

Ordinarily, to estimate the benefit-cost ratio satisfactorily, a number of subcriteria must be evaluated first. Subcriteria that make up total benefit and total cost for the proposed dam are apparent in the preceding ratio. In aggregate, these subcriteria determine the benefit-cost ratio.

2. Predicting how well each alternative will fare if it is adopted is the most demanding part of the decision-making process. For the proposed tire-mounter, the engineer had to predict the cost of building it, the number of man-hours needed to maintain it, its reliability, and so on. To make these predictions, the engineer relied mainly on his judgment and on experiments with a working version of his proposal. (The engineer has other means of predict-

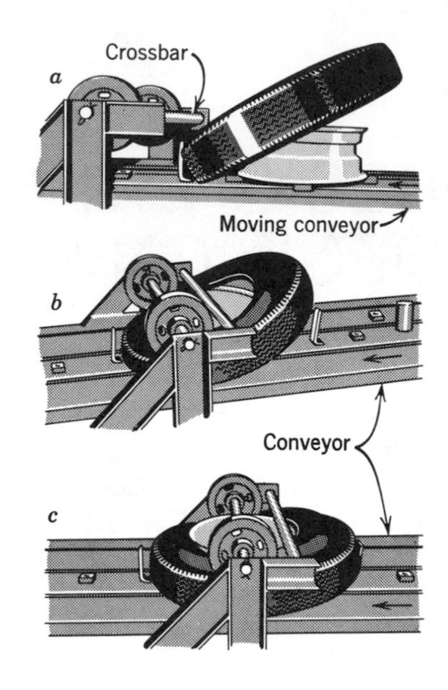

Figure 6-2 (a) *The beads of the tire are lubricated to insure that they will slip over the rim. The tire is on the wheel at an angle of approximately 30° to the rim as it approaches the cross bar.* (b) *As the tire moves under the cross bar it is engaged by two wheels located parallel to the centerline of the conveyor. These wheels are the only moving parts of the new machine; they squeeze the leading beads of the tire into the drop center of the wheel.* (c) *The remainder of the tire is forced onto the wheel by the cross bar.*

Criteria		Proposed Device	Present Machine
Investment	Cost to build, install, "debug"	$750	0
Operating expenses (for 5 years)	Attendant	$600	$23,000
	Maintenance	350	900
	Repair	250	750
	Power	0	400
	Total	$1200	$25,050
The unquantifiables	Reliability	Better	
	Safety	Better	

$$\frac{\text{Benefit}}{\text{Cost}} = \frac{\text{Saving in operating expenses}}{\text{Cost of building \& installing}} = \frac{\$25{,}050-\$1200}{\$750} = 31.8$$

Figure 6-3 *This is a page from the engineer's final report, comparing the performance of his device and that of the one presently used. He predicts that $750 will be required to build, install, and "debug" his device. The costs of an attendant, maintenance, and the like have also been predicted for the five years that he expects his device will be needed. He acquired equivalent data for the existing machine and summarized these costs in this table. He also calculated a benefit-cost ratio that certainly helps his superiors appreciate the merits of this $750 investment and you to understand why they adopted it so readily.*

ing performance, such as mathematics and simulation, which are described in Chapters 9 and 10.)

Predicted performances should be in the same units if they are to be accumulated and compared intelligently. By far the most convenient measure for these purposes is money. Thus, Figure 6-3 shows predicted performances in dollars for the criteria for which this is feasible. There will always be some criteria that defy numbers. Note that although it was too expensive and time-consuming to measure safety and reliability in this case, the engineer did not ignore them. He judged them in qualitative terms and concluded that they reinforced the monetary argument for this device.

3. To make an intelligent choice from the alternatives, they must be compared meaningfully with respect to the criteria. When dealing with criteria for which monetary predictions are feasible, these figures are usually tabulated or otherwise aggregated so that costs and benefits can be easily compared, as exemplified by Figure 6-3. Incidentally, this example, simplified

for our purposes, illustrates one of a number of methods of making economic comparisons of engineering alternatives. These procedures derive from a rather extensive body of knowledge of fundamental importance in engineering, generally referred to as engineering economics.

So much for steps in the decision-making process. The thoroughness and rigor with which these steps are executed vary considerably from problem to problem, depending on the complexity of the solutions being evaluated, their competitiveness, the importance of the decision, and other circumstances. In one situation the most elaborate, exhaustive procedures, involving much measurement, modeling, and cost investigation are warranted. In another, only a quick, simple, informal judgment is justified. The elaborateness of the evaluation process also depends on the stage of decision making. At the outset (widest part of the "diamond" of Figure 6-1), there are likely to be many competitors at least some of which, even though only vaguely conceived, can be eliminated by quick and simple evaluations. The remaining candidates, probably with details added, warrant more refined evaluations. As this multistage screening process continues, the competitiveness of the remaining candidates, the detail to which they are specified, and the elaborateness of the evaluations will increase.

Other than outlining this four-step procedure, there isn't much more that can be generalized about engineering decision-making procedure. Not so, however, about criteria employed in engineering decision making. There is a lot more to be said.

More on Criteria–the Elusive Ones

Only a minority of criteria can be simply and accurately measured. Most likely, in a given project, more than half of the criteria to be considered are either impossible or impractical to measure or, in some other respect, are troublesome for decision makers. Considering the difficulties these "troublemakers" cause, the risks associated with ignoring them, and the vulnerability of engineers in such matters, it is well worth discussing them at length. This discussion begins with a case in point.

A young engineer has been hired as a technical consultant to assist a city administration in the selection of a solid-waste management system. No doubt about it, there are alternatives —recycling, conversion to energy, and landfill, to name a few. In the course of his work for the city he naturally prepared a list of criteria on which the choice of system should be based. The criteria he proposed are:

1. *Initial investment.* **Purchase price of the land, buildings, and equipment.**
2. *Operating costs.* **Labor, energy, materials, and maintenance.**
3. *Environmental effects.* **Air, surface-water, groundwater, sound, and visual impacts.**
4. *Availability.* **It's important; it depends on frequency and duration of shutdowns for repairs, maintenance, and cleaning.**
5. *Safety.*
6. *Vulnerability to changing regulations.* **Governments have the nasty habit of periodically tightening the ground rules under which landfills, incinerators, and other solutions must be operated, and this often ups the cost of operation.**
7. *Public acceptance.*
8. *Adaptability to new materials.* **Alternative methods differ in their ability to accommodate new materials that are sure to come along. Landfill accepts everything, but others do not.**
9. *Predictability.* **There** ***are*** **varying degrees of uncertainty associated with different methods (e.g., the unpredictable markets and prices for recycled materials).**
10. *Expandability.*
11. *Skill required for operation.* **It's of consequence not only because of wages. Some methods require a greater percentage of high-skilled people, which yields not only an administrative headache but a different ratio of skilled to unskilled jobs on the public payroll.**
12. *Energy consumption.* **This is over and above the direct cost of energy required; here it is a matter of consuming versus conserving a societal resource.**
13. *Proportion of materials conserved.* **Like criterion 12, this is beyond immediate economic concern.**

Here are some observations I hope you made as you examined his criteria.

- **That the selection and weighting of criteria are consequential matters in an engineering project.**

- **That there are more criteria to be considered in such a decision than you surmised. The number in this instance is not unusual and, to be sure, it complicates things.**
- **That some criteria, like initial investment, can be evaluated in dollar terms and with relative ease.**
- **That others, like safety and public acceptance, are impossible or impractical to predict in numerical terms and therefore can only be considered qualitatively in the decision process. Engineers like to refer to these as the *unquantifiables* or *irreducibles*. These, indeed, complicate decision making.**
- **That there is considerable uncertainty inherent in the prediction of future legislation, new materials, and public acceptance. The result is that a surprising amount of "crystal balling" is required, which is always a source of a certain uneasiness.**
- **That a majority of these criteria involve consequences directly and only for the municipality using this solid waste management system. But here and in many other engineering undertakings *nonusers* are also affected. Remember, other communities, downwind and downstream, that do not benefit from this system still suffer the pollution consequences.**
- **That some of these criteria represent a long-term concern for the city's welfare. A criterion like expandability implies caring about possible expenditures perhaps 25 years hence.**
- **That some criteria (e.g., energy consumed and material resources destroyed) require a concern that extends beyond city limits and beyond the immediate future. They imply caring about the material and energy resources available to future generations.**
- **That all this adds up to a challenging decision-making situation because of the number of criteria, the conflicts between them, the predominance of unquantifiables, and the uncertainties involved.**

This engineer did his job well. His list of criteria reflects no neglect of unquantifiable criteria, no disregard for indirect ef-

Figure 6-4 *Aesthetically, how do you rate it? This and other stations in the San Francisco Bay Area rapid transit system are generally conceded to excel in this respect. Aesthetically, the vehicles–inside and out–as well as the building, appeal to almost all (never expect all) users.*

fects, no blindness to the consequences for nonusers, and no indifference to long-term implications. And that's the way it should be.

Aesthetic Quality

The ultimate in defiance of quantification is the visual appeal of a design—its aesthetic quality. This is a criterion that is almost as difficult to verbalize as it is to measure. If you are pressed to express your reaction to the structure of Figure 6-4, you are likely to respond with "striking" or "neutral" or "ugly" or some other expression of visual impact. The next person, viewing the same scene, might well respond quite differently, and the next still differently. This is natural, since the aesthetic quality of a crea-

Figure 6-5 *How do you rate this one? In contrast to the designs pictured by Figure 6-4, this water tower is controversial with respect to visual appeal.*

tion is the manner in which the viewer perceives it; beauty is in the eyes of the beholder, according to an old saying.

Consequently, aesthetic quality is the most troublesome kind of criterion; it requires that the designer anticipate the reactions of others, many others, in the face of widely varying individual responses. His final design (e.g., Figure 6-5) is what he anticipates will please the most viewers of his creation, and it may or may not coincide with his personal likes and dislikes. Observing what a challenge this is, you can readily surmise why some engineers, architects, and others in the design professions dodge this matter of aesthetics, *and* why authors on the subject of design do likewise. Yet this is an area in which engineers are frequently criticized. So apparently the tendency to slight aesthetic considerations is hurting our creations and our image. As elusive as it may be, this criterion is worthy of your attention, starting now.

Usability

How easy is the oven to operate, the automobile to repair, the production machine to stop in case of emergency, the fire truck to set up for action, the highway signs to follow, the computer to use, the camera to load? These questions call attention to the need for carefully considering the ultimate users of an engineer's creations—the people who must operate or otherwise use them. There are very few (can you think of any?) engineering creations with which humans don't come in contact in some capacity—users, operators, or maintainers—so that a criterion to be considered in almost every problem is usability (some prefer operability).

I have a thick folder of clippings reporting designs that are blunders from the user's point of view. One involves a movie projector with the two switches shown here. That seems harmless, except the instruction book emphatically cautions, "DO NOT REVERSE DIRECTION WHILE PROJECTOR IS ON." Surely *that* is going to happen more than once. What makes this example worth mentioning is the remarkably simple way these ON-OFF and reversing functions can be combined into one switch, which eliminates the possibility of reversing with the motor on. Other examples are: an automobile with the steering column placed so that the driver's foot is easily blocked as he tries to shift it from accelerator to brake; the highway interchange that seems to invite drivers to enter by an exit; truck cabs outrageously awkward to get in and out of; hazardous appliance plugs; and lecture hall controls in ridiculous locations. Surely you have encountered similar cases; everyone seems to have their pet gripes in this respect. Who is to blame? Engineers, architects, industrial designers, and automobile stylists who simply haven't given enough attention to the people who ultimately will use—or try to use—their creations.

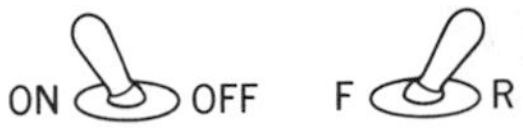

ON OFF F R

In a majority of instances designers have done their job well in this respect, and sometimes exceptionally well, as illustrated by Figure 6-6. Or sometime, if you have the opportunity, take a look at the shape-coded controls and other considerations of the user in the cockpit of a commercial airliner. (A lot of attention has been given to preventing pilot goofs by the design of his instruments and controls.) And I notice that the designers of my pocket calculator are thoughtful of users. For instance, the tilt of the display improves visibility; the click of the function keys assures me that the function has registered; and the ON-OFF button is located to minimize the chances of accidentally switching the machine on when sliding it into a case or pocket. How do the designers of graders, aircraft, calculators, and a majority of other engineering contrivances avoid blundering in their consideration of the user?

Figure 6-6 *To score well with respect to the criterion usability, a machine like this grader must provide for good operator visibility, controls that are easy to learn and readily accessible, a seating arrangement that minimizes operator fatigue, and a variety of other operator-oriented features. And this one does; the designer has done an excellent job in these respects.*

For one thing, they devote a lot of forethought to the user —what he needs, his habits, and his physical, sensory, and mental limitations. Furthermore, they frequently conduct field trials with prototypes to evaluate usability. In addition, they request help when they need expert advice in such matters, which they can get from handbooks and specialists in the field of *human engineering* (some prefer the term "human factors").* Finally, they apply a lot of common sense and thereby avoid the design blunders you read about in *Consumer Reports*.

* Morgan et al., *Human Engineering Guide to Equipment Design*, McGraw-Hill, 1963. This is one of a number of sources of information concerning design for the user.

The tendency of some practitioners in the design professions to slight the criterion usability is puzzling to me, partly because so much of this is common sense, and partly because such considerations are the most interesting part of my work as an engineer.

Safety

Certainly, of the alternative schemes considered in the design of an improved tire mounter, some were more likely to injure workers than others, and this *does* matter. Although the designer did not attempt to *measure* this characteristic of alternatives, this in no way reflects the importance he attached to the safety criterion. The fact is, safety of *proposed* solutions is virtually unquantifiable. As elusive as it is, however, this is a criterion engineers cannot afford to slight. A mass of consumer protection legislation and a dramatic increase in liability for product-caused mishaps make thorough consideration of safety a must.

In order to fulfill his professional obligations with respect to safety, an engineer must:

- **Avoid design blunders, such as miscalculation of the shear force on a strategic load-bearing pin in a piece of agricultural machinery, causing it to fail in use.**
- **Anticipate how his creations *will* be used in the field, not simply how they *should* be used. (Manufacturers are liable for product-caused accidents even when the injuries arise from misuses of the product.)**
- **Strive to foresee all of the possibilities, albeit out of carelessness, for injury to the user, *and* to accident-proof his creations accordingly (e.g., through hand guards, electrical interlocks, overload sensors, and emergency switches).**
- **Exhaustively test his solutions and strategic components thereof, in the laboratory and in the field, for vulnerability to hazardous failures (e.g., an axle prone to break because of metal fatigue).**
- **Remain informed about the numerous codes and standards prescribed by law, professional and industrial organizations, Underwriters' Laboratories, and the like.**

And this is not all that safety consciousness implies for the engineer, but this sampling suffices to demonstrate that there is a lot to this matter.

Product safety considerations present a special challenge, and not only because of their importance. Some of this challenge derives from the quantification difficulty. Furthermore, this matter of foreseeing all the accident-causing eventualities in the manufacture, use, and servicing of a product certainly doesn't make things simple either. Take this case. An automobile company was sued (and subsequently paid) because a front brake hose burst, causing an accident. It burst because a front wheel had been rubbing it on turns, and this rubbing was traced to sloppy assembly and inspection at the factory. The prevailing attitude of the courts today is that this rubbing and subsequent failure because of improper assembly and careless inspection is foreseeable, so it behooves the engineer to anticipate such unlikely chains of events and, if possible, foolproof his designs accordingly. Reading the case histories of product liability suits is a journey into the bizarre, and although you might question the foreseeability of many of the accidents, the courts don't.

Some Effects of Increasing Complexity

Some of the criteria that come up repeatedly are increasing in importance as engineering creations become more complex. One of these is *reliability* which, by the way, has a very specific meaning to an engineer. To him, reliability is the probability that the component or system in question will not fail during a specified period under prescribed conditions. For example, the reliability of video tubes made by the Zeon Company is 0.95 for the first year of operation under "normal" usage, vibration, and temperature conditions. Reliability is especially important when failure is costly, which is true of an amplifier in a transatlantic cable under 3000 meters of water or in an orbiting communications satellite. It is also important when people are highly dependent on the system, which is true of public water supply systems, defense missiles, and elevators in a 40-story building.

Reliability, along with criteria such as repairability, maintainability, availability, and usability, are more important not only as a result of complexity; the heavy investments required by many modern engineering creations and our dependence on them also add weight to these criteria. A whole production system depends on that simple little tire mounter; a corporation can "go to pieces'' if its computer breaks down for long; and we all know what happens when a city's electric power system cuts out for a few hours!

Note that specific criteria like reliability and repairability and, ultimately, the total cost of a solution, can be improved through simplification. Simpler mechanisms, simpler circuits, simpler functioning, simpler shapes, and so forth are generally worth striving for. This fact, combined with the seemingly irreversible trend toward greater complexity of man's creations, prompts this special plea on behalf of the simplicity criterion.

A Plea for Simplicity in Design

Usually, among the many solutions to a problem, some are outrageously complicated; others are refreshingly simple but no less effective than their complex counterparts. The tire mounter provides an excellent example. The previous machine was unnecessarily complicated; it had 60 moving parts and required electricity and compressed air. The new device is a sharp contrast: it has only a few moving parts, requires electricity only, and there is practically nothing to fail or maintain. *This* comes from good engineering.

Another example: an artificial satellite tends to tumble indefinitely as it orbits Earth, yet for certain applications one side of the satellite must remain pointed toward Earth at all times. This is true of the weather satellite; its cameras will hardly do us any good if they are pointing out into space. Therefore, such a satellite must have an orientation system that prevents it from tumbling and keeps it properly aimed. A long boom attached to the satellite provides a beautifully simple solution to this imposing problem, as explained by Figure 6-7.

To fully appreciate the virtues of this solution, you must know something about the alternatives. You would be amazed at the complicated systems that have been used or have been proposed to achieve this same purpose. One system uses electronic horizon sensors in conjunction with a system of gas jets, which means that it requires compressed gas and electricity (the gravity gradient method requires no power). In addition, it has about 40 times as many parts, weighs about four times as much, costs about 50 times more, takes more space, and is considerably less reliable than the gravity-gradient system. These are no small matters.

The stabilizing boom on a satellite may be up to several hundred meters long, and it obviously cannot be protruding from

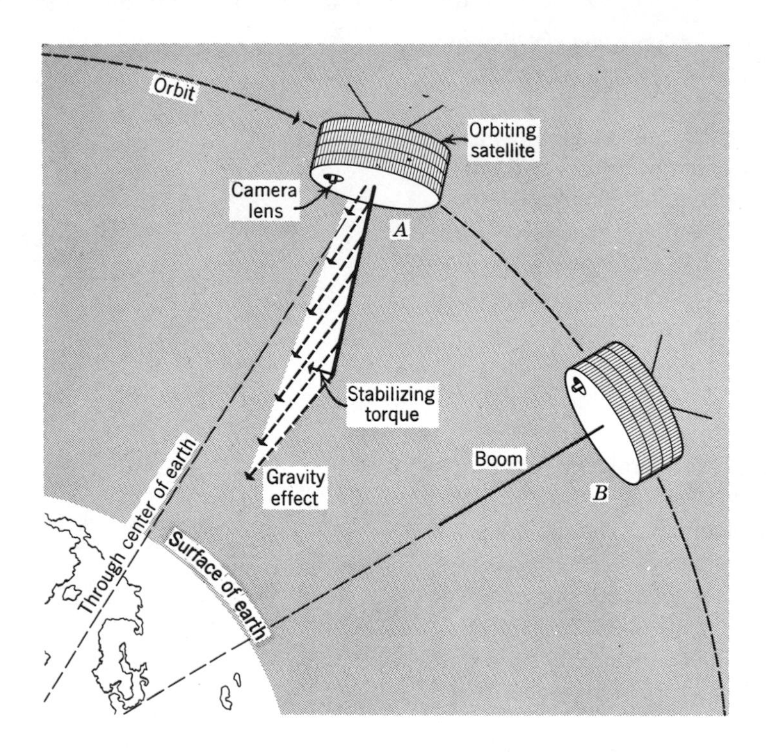

Figure 6-7 *The pull of gravity on the end of the boom closest to the earth is greatest at that point (the force of gravity varies with the square of the distance to the earth's center). This swings that end toward an imaginary axis extending from the satellite to the earth's center. This gravity-induced torque causes the structure to seek the desired vertical orientation. In position* A, *gravity is in the process of orienting the still-tumbling satellite. Eventually the structure reaches and remains in orientation* B. *In space jargon this is referred to as the gravity-gradient method of satellite stabilization. (Incidentally, since there is no air resistance, this structure will act like a pendulum on a pivot and swing to and fro indefinitely unless some provision is made to dampen this oscillation. This is done, for example, by attaching a tiny weight to the tip of the boom by means of a coil spring, so that the weight flops to and fro out of phase with the boom and eventually kills the oscillations. It is possible that by chance the satellite-boom structure will orient itself with the satellite facing the wrong way, which means that the satellite's cameras would be forever aimed at outer space. In this event, the boom is retracted, the satellite is allowed to tumble, then the boom is reextended, and this process is repeated until by chance the structure assumes the desired orientation.)*

the satellite during the launch phase. So how does a satellite "grow a boom," say 25 meters long, after it is in orbit? Many elaborate schemes were considered, but the solution eventually conceived is remarkably simple (see Figure 6-8), thanks to an engineer's conviction that there must be a better way.

Notice some of the virtues of this simple "boom-grower," especially in contrast to the solution that might have evolved. It is compact, light, reliable, relatively inexpensive, and requires very little power. In fact, the spring action of the metal provides the energy needed, so that it can be operated without a motor. Engineers call this an *elegant* solution because it achieves the desired objective in an uncomplicated manner.

Will *you* recognize design simplicity as well as the lack of it? Will you strive for it? As an engineer, you should not be content until you have simplified to the maximum possible extent the mechanisms, circuits, method of operation, maintenance procedures, and other features of your solutions. Ordinarily there is a big difference between a workable solution at the time it is conceived and that solution after it has been effectively simplified.

The simplicity of the adopted solutions in these three cases is impressive. However, simple designs, per se, are valuable only in the satisfaction (don't underrate it) that they bring to their creators. The payoff comes from the economy of manufacture, low maintenance cost, and high reliability that results from simplicity.

Observe how a cleverly simplified solution tends to deceive the untrained eye. It belies the difficulty of the problem and all the knowledge, skill, and effort that went into its solution. This is certainly true of the examples just cited. How easy it is to grossly underestimate what underlies this simplicity, especially if you are unaware of the overly complex solutions that could have evolved (and sometimes do). When an engineer's solution has reached this deceptively simple state, he can consider his job done well.

A mark of an exceptionally creative person in almost any field is the simplicity of his creations. Observe the few lines required by a good cartoonist to create the desired effect, the remarkably few brush strokes evident on a good watercolor, the few well-chosen words required by the skillful author, the simple lines that make the great works of architecture, or the striking simplicity of the three engineering creations just described. Yes, there is art in engineering and, by the simplicity criterion, these solutions are exceptionally artistic.

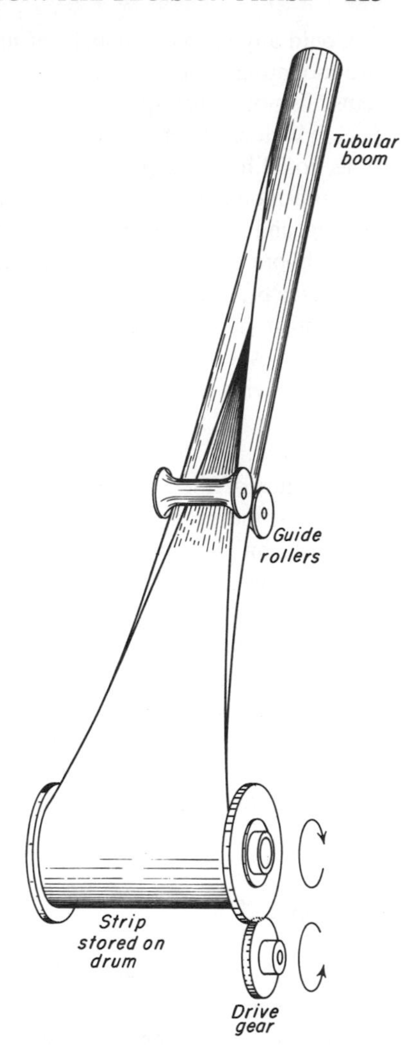

Figure 6-8 *Design concept for "growing" a long tubular boom on an orbiting satellite. A flat metal strip the length of the desired boom is treated so that its normal form is tubular. Then it is rolled up flat onto a spool and stored in the satellite. After the satellite is in orbit, a signal actuates a motor that unrolls the strip. As it unfurls, the strip returns to its tubular form and, when fully extended, is the desired boom. Tubes hundreds of meters long can be so constructed.*

Could anyone have thought of the solutions described for the tire-mounting, boom-growing, and satellite stabilization problems? If not, what special knowledge was required? A clever person without a technical education but with mechanical inclinations could have thought of the solution concept underlying the new tire-mounter, but he would have been hard put to refine his concept and to specify the details of a workable device. In the case of the boom-grower, a good machinist could have come up with the concept, but the engineer who developed a functioning mechanism had to have special knowledge of metals, heat treating, stresses, mechanisms, and cantilever beams. Only a person familiar with the principles of mechanics, gravitational phenomena, oscillatory systems, and the like would be equipped to conceive, let alone develop the gravity-gradient idea.

The nub of the solution in each of these instances is pure invention, the product of an engineer's ingenuity. But without special technical knowledge, it would have been virtually impossible to convert the basic idea to a workable solution in any of these cases. So although invention is a necessary and very important part of engineering, it is hardly sufficient.

This analysis highlights a major difference between present engineers and engineers of preceding periods like Edison, Whitney, and Watt. All are inventors, but there is a substantial difference in the amounts and types of knowledge available to engineers of the past and those of the present.

EXERCISES

1. List the criteria you believe should be considered by designers of one of the following: a highway, an artificial hand, a method of assembling automobiles, or the automobile itself.

2. Assume that the memorandum of Figure 6-9 (or one supplied by your instructor) is directed to you. Thus it is your job to prepare the preliminary report that Chief Engineer Hogan requests. The primary purpose of this assignment is to give you an opportunity to practice what you have read about design technique in Chapters 4 to 6. So, in addition to a brief report on your solution, in an appendix to that report, your instructor will expect ample indication that you tried your hand at the design procedure described herein. Cite your formulation and why you chose it, a problem analysis, the many alternative solutions you considered,

DESIGNWELL ASSOCIATES

Consulting Engineers Inter Office Memo

TO: Stu Dent, Project Engineer

FROM: James C. Hogan, Chief Engineer

This is to verify our conversation of yesterday. The Red Cross seeks to develop a low-cost disaster-victim shelter. They have asked us to submit preliminary plans for a system that meets their specifications.

They have in mind an air-transportable shelter weighing not more than 12 kilograms, occupying no more than one-tenth of a cubic meter unassembled, and suitable for 4 persons. They want these structures to be readily interconnectable to accomodate larger social units.

Remember, it will be assembled by unskilled persons, many of whom cannot read.

The Red Cross plans to stock 500,000 of these units, distributed over a half-dozen sites around the world. For fast deployment, these units should be stored for quick loading and unloading.

I am asking you to prepare a preliminary design and to get your report to me in the very near future.

Figure 6-9

and so on. You are not expected to produce a "drawing board solution" to this problem; a design concept, presented in words, sketches, and diagrams, will suffice. (This does not mean, however, that you can leave most features of your proposal to the instructor's imagination!)

3. Continue with exercise 3 (animal lift problem), or 4 (trash sorting), or 5 (iceberg), or 6 (wrong way), at the end of Chapter 5, following the general instructions given in exercise 2 above.

4. Can you picture it? You are an engineering consultant to a grower of thousands of acres of tomatoes. Your mission is to develop a machine that will significantly reduce harvesting cost.

You now have preliminary designs. They are: (a) machine-assisted handpicking, and (b) completely mechanized harvesting which, of course, requires a larger investment but has a lower operating cost than alternative *a*. What criteria are you going to apply in choosing between these? How do you suspect the alternatives will fare with respect to each of these criteria?

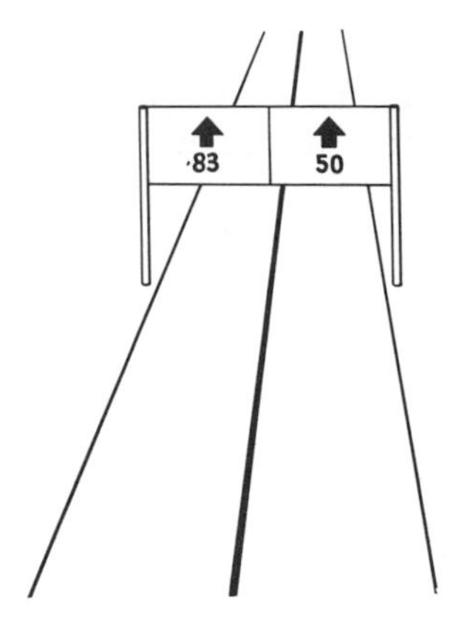

5. Many a driver has jockeyed at 95 kilometers per hour to get into the left lane only to discover a half-mile later that the message this sign was meant to convey was, "Be in right lane if you intend to turn right on Route 50." Surely *you* can tell highway engineers how to accomplish this, and simply at that!

6. What is the simple remedy to the projector control problem described on page 109?

7. Choose a consumer product, such as a kitchen stove, automobile, or camera, and evaluate the job the designer did from the usability viewpoint.

DESIGN: SPECIFICATION AND WHAT FOLLOWS

Chapter 7

At the conclusion of the decision phase, the preferred alternative is in the form of a solution concept (some call it design concept). This is the essence, the guts, the major features and operating principles of the solution, with details like particular materials, exact dimensions, and the like still lacking; hardly in presentable shape! Thus the development of a solution concept is not the end of an engineer's encounter with a problem. A solution in the form of pages of rough sketches, notes, and computations, with many details the designer has yet to put on paper, is unsatisfactory for several reasons.

One reason, and few engineers can escape it, is that they do not have the final word on implementation of their solutions. Almost invariably there are others above them in the decision-making hierarchy whose approval they must gain. Although acceptance was virtually automatic, the designer of the tire-mounter had to get a green light from his immediate supervisor, who is the manager of engineering, *and* from the superintendent of manufacturing. Failure to win the backing of either would have killed his proposal.

And let's face it, the people who can veto an engineer's works are usually hard-nosed businessmen or tightly budgeted public officials who must be convinced. To do so, he must be quite specific about his proposals as well as their virtues and limitations. And so the designer must describe in considerable detail the physical attributes and performance characteristics of what he has in mind.

There is another and very practical reason for this. An engineer who designs a machine or structure never actually builds, operates, or services his creation. This makes it essential that he amply document and effectively communicate his proposal to the persons charged with construction, operation, and servicing so that they can satisfactorily fulfill their responsibilities.

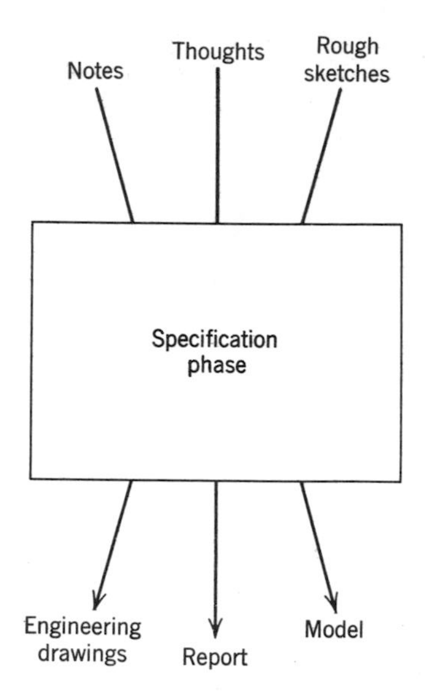

Specification Phase of the Design Process

The upshot of all this is the specification phase of the design process which, among other things, ordinarily yields detailed, dimensioned drawings of the final solution. These are usually referred to as engineering drawings (sometimes as the prints) and, for a complex system like a bridge or chemical plant, they may number in the thousands.

Another medium for communication of the designer's solution is the engineering report, usually a rather formal and voluminous document describing and evaluating his proposal especially in words. Many a fledgling engineer finally comes to appreciate the importance of verbal skills when he is preparing his first report of this kind. (*You* needn't wait that long.)

Occasionally he supplements the prints and report with a model, which may be a working model for testing and demonstration purposes. And, finally, his communication efforts may also include an oral presentation of his proposal to superiors, clients, and sometimes the public. That, too, calls for a skill worth cultivating.

The specification phase is likely to involve a lot of detail work. Draftsmen and other technical assistants relieve him of some of this burden but, in general, the engineer must specify the types and properties of materials to be used, exact dimensions and tolerances, methods of fastening, operating temperatures, voltages, and so on. For obvious reasons, this seldom proves to be the most enjoyable part of his work.

The Design Cycle

The delivery of a final report terminates the engineer's design efforts on a given project but certainly not his responsibilities. The professional extends himself to gain acceptance of his design, to oversee its installation and use, and to accomplish other postspecification functions, identified in Figure 7-1. This cycle of events is worth examining.

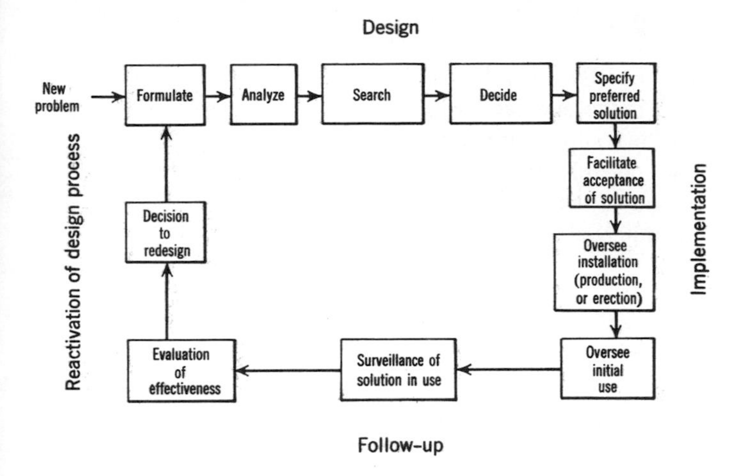

Figure 7-1 *The design cycle.*

Implementing A Solution

Never assume that a solution will automatically be adopted, properly constructed, and used as intended. Many things can go wrong, and measures must be taken between the time a solution is proposed and the time it is an accomplished fact to prevent them from doing so.

For example, measures are needed to ensure that a solution will be adopted by the appropriate people. Engineers often begin their careers with the mistaken impression that, if their proposals are technically and economically superior, they will naturally be implemented. But engineering is ordinarily a staff function in an organization, so that engineers issue recommendations, not commands. This, plus the fact that differences of opinion do arise, makes it imperative that careful attention be given to this matter of gaining solution acceptance.

Young engineers are inclined to become discouraged if several of their proposals have been rejected. They are inclined to blame other people, the organization, anything or anyone but themselves. But the chances are that they have underestimated the need for effective presentation of their proposals, for convincing others on the worth of their ideas, for a certain amount of realistic compromise, and for careful planning to minimize resistance to change.

Follow-Up

Periodic monitoring of solutions in use is especially valuable as a means of improving future designs. Rare is the engineer who cannot benefit by watching people operate the computer terminal he designed, service his automobile engine, or follow his highway directional signs. Not only is this kind of activity educational, but it is likely to be one of the more interesting aspects of his work.

Reactivation of the Design Process

Periodic evaluation of solutions in use also provides a basis for deciding when to redesign. No solution to a practical problem remains superior indefinitely. Better methods are discovered, new demands arise, new knowledge accumulates, conditions change, and physical depreciation occurs. Consequently, a point is reached in the life of a design at which it is profitable to seek a better solution. An engineering department can intelligently decide when to engage in redesign only if the current solutions to problems within its realm are periodically appraised.

The design cycle is complete when, after a solution to a problem has been devised and used over a period of years, it appears that redesign will prove profitable, and the process of designing a superior solution is again initiated.

APPLICABILITY OF THE DESIGN PROCESS

What you have been reading about in the past three chapters can be applied to a broad range of technical and nontechnical problems.

Technical Applications

You had a look at the "boom-grower" in Chapter 6. Here is some insight into the birth of that mechanism. An engineer was asked to find a means of constructing a 25-meter boom, 5 centimeters in diameter, solid or tubular, on a 100 × 60 centimeter hatbox satellite *after* it is in orbit. Figure 7-2 indicates what he did first—set up the problem.

When he was satisfied with his definition, he went to work on alternative solutions. Figure 7-3 shows some of the possibilities

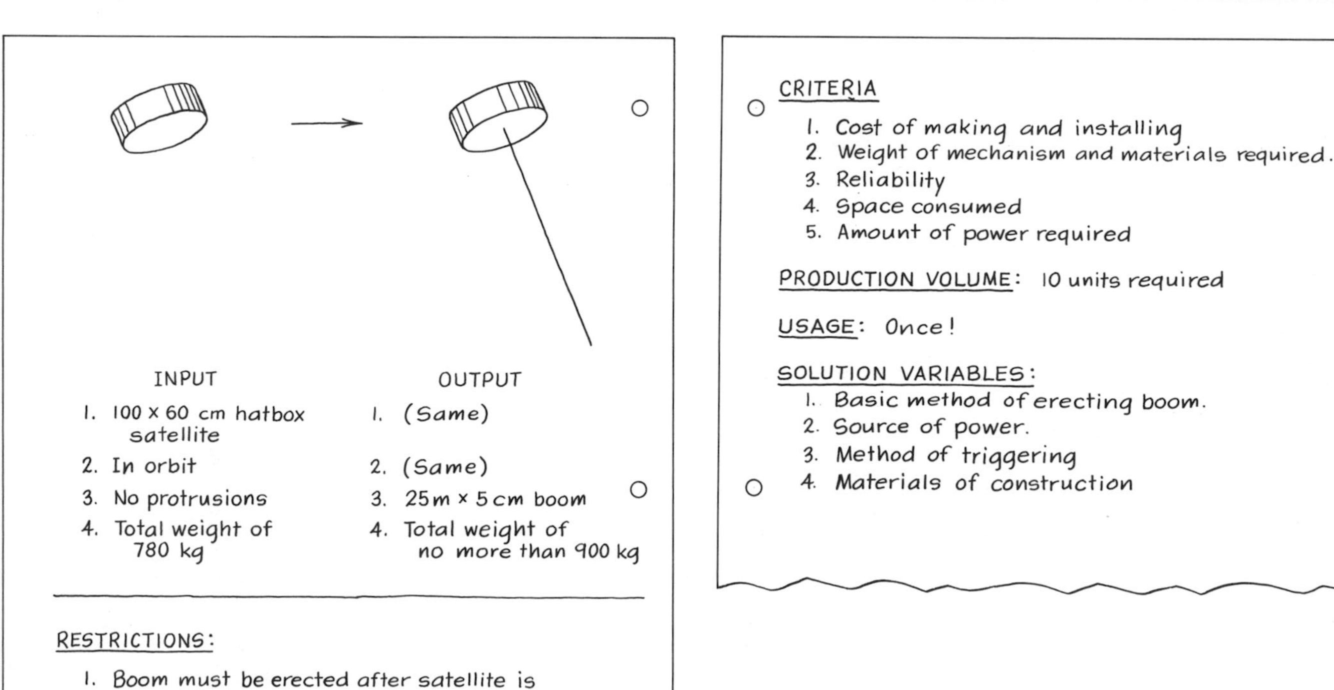

Figure 7-2 *These two pages from the engineer's notebook indicate how he defined his problem. (I have eliminated some of the details and simplified others.)*

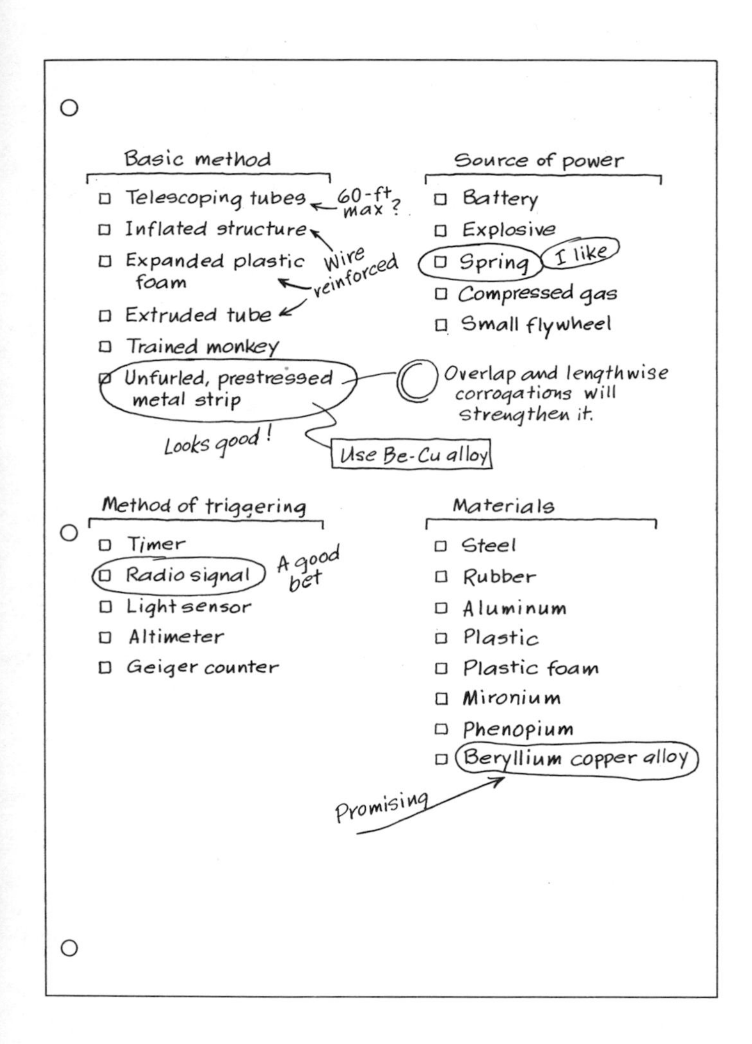

Figure 7-3 *Another page from his notebook (again, I have done some simplifying), showing some of the alternative partial-solutions that he accumulated for the major solution variables. Based on what I said in Chapter 5, you are not surprised that I am happy with the manner in which he used solution variables to guide his search. The thoroughness with which he capitalized on this approach may well account for the elegance of his final solution.*

he came up with. To him, in retrospect, and probably to you now, Figure 7-3 makes the whole thing look simple and obvious. But what you see in this figure is the product of about three days, not all at one stretch, of thinking, talking, reading, visiting, corresponding, telephoning, sketching, daydreaming, sweating, scratching, smoking. . . .

What followed was the most time-consuming part of the task: gathering performance data and cost figures, adding details and more details, calculating, experimenting, comparing alternatives, and so on, until he had tentatively decided on the mechanism pictured in Figure 6-6. A working model of that concept was constructed and tested, after which minor changes were made. He continued this test-and-modify cycle until he was satisfied. The rest of the job was a matter of preparing detailed, dimensioned drawings and his final report.

Those last few sentences make the task of converting concept to reality sound so simple and routine; some of it, yes, but much of it—no. Here is a brief glimpse at the kinds of things he ran into. The extended boom is subject to several types of forces, the magnitudes of which must be predicted. One of them is a bending action generated by sunrays striking one side of the boom (something that many of us could easily have overlooked). The challenge here is to predict the heat-generated forces in the boom, then do the same for other forces, and then to determine the structural shape and material thickness necessary to withstand the resultant of all forces. This is only one of hundreds of details that must be resolved to convert the original design concept to a fully specified, workable solution, details the layman quite naturally overlooks as he marvels at the remarkably simple result.

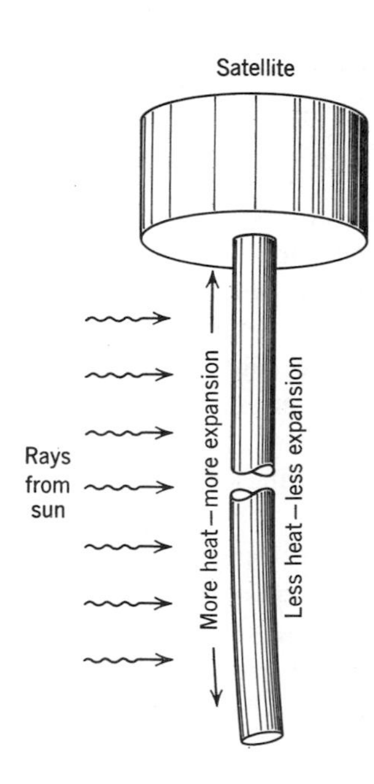

The preceding case is a beautiful application of the design process to a technical problem. *The same basic process applies regardless of the nature of the problem.* To illustrate this broad applicability, here is an elaboration on the spectrum of activities bounded by development engineering and sales engineering, which was introduced in Chapter 3.

Task A, illustrating development engineering, involves design of a device that will directly convert the spoken word to printed form. The input is an oral message, the output must be a record of that message on paper.

Task B, representative of sales engineering, is at a company that manufactures electrical equipment (motors, transformers, etc.) and assembles these components into power systems designed to suit the unique needs of individual customers. Most of these customers are factories, refineries, printing plants, and the like. Task B involves calling on a prospective customer, in this case a paper-manufacturer who plans to construct a new plant. At the potential customer's invitation, the engineer familiarizes himself with the paper-making process, makes a thorough investigation of the company's needs, and then designs a power system adapted to the customer's process. In doing so, he relies heavily on components manufactured by his employer. He submits his solution along with the price to the potential customer. If his system is purchased, he will oversee its installation and remain in contact with the job until it is running smoothly.

Task A is concerned primarily with the *conception* of a new device. There is a minimum of past experience on which to rely; much original thought is required; some research may be necessary; the work can get steep technically.

Task B is mainly the *application* of existing devices and components in the satisfaction of the unique needs of specific customers. For this type of work the engineer can rely on a large backlog of experience in designing such systems in similar situations. Although each situation he encounters is different to a degree, the work involved is hardly what you would consider pioneering. Whereas the main challenge of Task A stems from the original nature of the work, the main challenge of Task B is in acquiring a thorough understanding of the needs of the customer.

These two tasks lie close to the extremes of a wide range of types of problem-solving activity. Along this spectrum lie such tasks as the design of an interplanetary transport, artificial heart, automobile, cement plant, bridge, computer, pipeline, undersea vessel, you name it. In all cases there is a problem to be defined in terms of a transformation, criteria, restrictions, and so forth; there are alternative solutions to be identified, choices to be made, a proposal to be specified.

$$3x + 2y = 12$$
$$2x + 7y = 25$$
$$x = ?$$
$$y = ?$$

⇩

$$3x + 2y = 12$$
$$2x + 7y = 25$$
$$x = 2$$
$$y = 3$$

Nontechnical Applications

Picture yourself with a set of simultaneous equations to solve. A transformation is involved (and although it is spelled out here, in practice this would certainly not be necessary for a problem so obvious). To achieve this transformation, you should not automatically resort to a particular method. There *are* alternative solutions, such as the substitution method, graphical means, determinants, or computer. (If the use of the word "solutions" here bothers you, recall that solution was defined as "a method of achieving the desired transformation." This may well run contrary to your customary use of "the solution" as well as of "the answer.") These methods differ with respect to criteria like time required, equipment needed, and accuracy attainable. Naturally, you will apply these in selecting the best method for the particular situation at hand. Furthermore, your teacher has imposed at least one restriction—that *you* solve it. Thus, this and other mathematical problems can be structured and approached according to the pattern introduced in preceding chapters, with some benefit to you *if* you are willing to give the proper emphasis to alternatives and criteria.

The transformation involved in a physician's typical problem is obvious enough. He works under restrictions set by statute, professional code, and sometimes the patient. He has alternative solutions available in almost every case; for instance, for ulcers, the main alternatives are diet, surgery, radiation, and medication. Certainly he applies a number of criteria in selecting the best cure. Among them are probability of success, time required, cost, and discomfort.

A course is intended to transform certain qualities of students. Suppose that you have been assigned to develop an introductory course in engineering. Surely, in defining this problem, you would thoroughly specify the knowledge, skills, and attitudes you would want to find in students when they had completed this course. Similarly, for the student "input," you would learn the entering students' backgrounds, misconceptions, apprehensions, and other facts. With this specific information on "input" and "output", you would know the transformation that your course must bring about.

Ordinarily, restrictions on courses are imposed by accrediting agencies, college regulations, and the like. In this instance, you are told that it must be a one-term course.

Among the criteria you would employ are benefit to the student, appeal to him, student hours required, resources required, and development cost.

In this problem, too, there are solution variables, such as method of communicating to the student and means of obtaining feedback. For each of these there are numerous alternatives.

Since mathematical problems, medical problems, educational problems, business problems, you-name-it problems all involve transformations, restrictions, criteria, and so forth, they are *basi-*

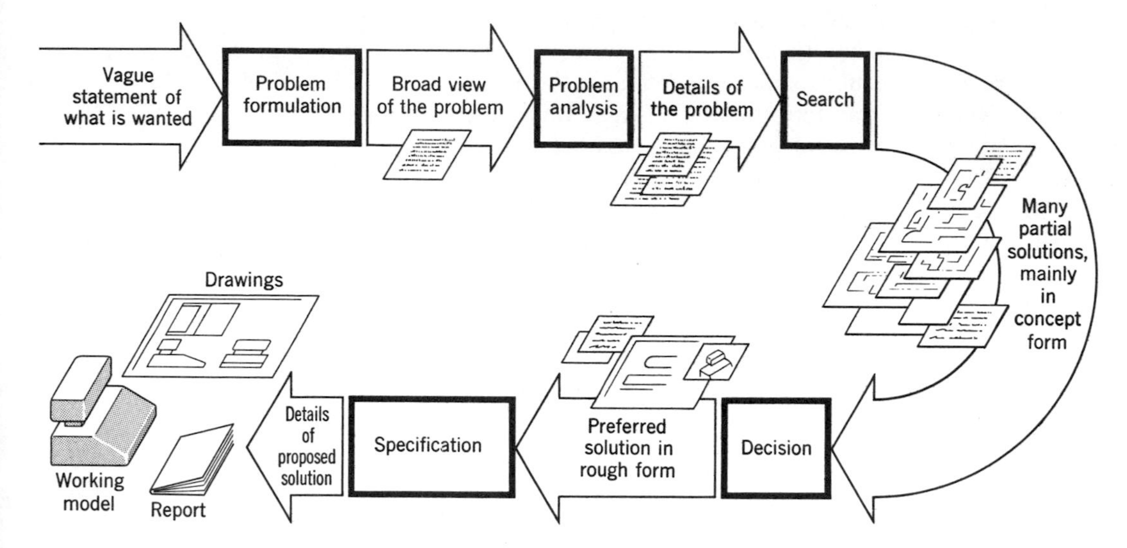

Figure 7-4 *The design process, picturing the input and output of each phase.*

cally no different from what, up to this point, have been called technical problems. It is conceivable then that you have learned something about problem solving that you can apply personally as well as professionally.

Now don't get me wrong; I am not suggesting that you formally structure every problem that you come upon along these lines, or that you always engage in an intensive search for solutions. I don't recommend this; I don't do it myself.

THE DESIGN PROCESS —AN OVERVIEW

Clearly this series of chapters deals with the principal activity in engineering—problem solving. However, in the conversation and literature of engineers, this activity is often referred to as design, or engineering design, or the process of design, or the design process. Such terminology is not standard and, furthermore, interpretations of a given term vary among users. In view of this, let there be no doubt about the preferred term in this book—*design process*.

As used herein, the design process is what Chapters 4 to 7 are all about. It begins with a vague statement of objective, ends with the specification of a solution, and embraces all activities and events between. As pictured in Figure 7-4, this process is a series of stages in the evolution of a solution. The purpose of each phase is different; the type of problem-solving activity that predominates in each is different. However, these stages have no sharply defined boundaries nor are they the orderly progression of distinct steps the idealist would hope for. There is a certain fuzziness as the emphasis shifts from one phase to the next. Occasionally solutions will occur to the designer while he is defining his problem or, during the search phase, he may decide to reformulate the problem. Similarly, it is impossible not to do some evaluating during the search phase. Chance plays a significant role in this process; new information and new ideas are uncovered unpredictably; blind alleys are encountered; occasional backtracking is inevitable. The process is clearly and unavoidably iterative.

Practicing What You Have Been Reading

With the temptations, pressures, and pitfalls that face you when you graduate from the classroom (the subjects of next chapter), it behooves you to be prepared by perfecting your design skills while in college. One consequence of being unprepared in this respect is that you surely will be victimized by certain habits of thought that hurt your performance on the job, such as the temptation to bypass problem definition. If you are to overcome such habits, you must work at it. That means reading and hearing about how to do it and, since this is a skill, it also means practice. While practicing, evaluate your problem-solving approach and discipline your mind accordingly. Such efforts will pay off in superior design performance over the years.

The Recurring Question of Economic Feasibility

There are remarkably few things that man cannot achieve given sufficient time and money. Whether a device *can* be developed for a given purpose is seldom the question. It can almost certainly be done if someone is willing to pay the price. The real question usually is, can a solution be developed profitably? Therefore, in most instances, technical feasibility is not an issue; economic feasibility definitely is.

Implicit in the inception of a design project is the *assumption* that a solution to the problem is economically feasible, that investment of engineering and other resources to develop a solution will be repaid with a profit. Before starting to design a machine for assembling reed switches, the engineers deliberated over the prospects of a profitable result. The decision to proceed was based on their judgment, on past experience in similar ventures, and on a willingness to accept a certain risk.

From the moment a project is begun, the hypothesis that a profitable solution will evolve is under test. This question is being asked repeatedly, although not always explicitly, through the design process: "On the basis of what has been learned thus far in the project, are the prospects of the venture yielding a satisfactory return on what apparently must be invested sufficiently high to justify continuing?" Thus, at any time from its inception to the specification stage, a project is subject to termination if the information accumulated indicates that a profitable solution probably will not be found under the current state of technology.

Of course, at the outset of a design project, relatively little is known about the probable final solution, so that at this stage there are many uncertainties and thus a sizable risk of being mistaken in assuming that the venture is economically feasible. As the project progresses, alternative solutions evolve, possibilities are appraised, and other information on which to base an answer to the ever-present profitability question becomes more substantial. Therefore, the risk of making the wrong decision is maximum at the start of a project and decreases thereafter as progress is made and evidence accumulates.

Making recommendations concerning economic feasibility of proposed projects is a very important part of an engineer's work. Such decisions are far from simple, and yet the engineer who acquires a record of blunders in this respect is not favorably regarded.

Terminology Note

Of late there has been a flood of books and articles proclaiming the virtues of a "new" problem-solving approach. As a consequence, sociologists, psychologists, business executives, economists, hospital administrators, government people, and all readers of the help wanted ads in the big city papers are talking about *systems analysis*. However quick writers may be to use the term and speak of its virtues, they are just as slow to tell us what it is. As it turns out, there are numerous interpretations. Some authors define systems analysis literally. To them, it is simply the process of identifying subsystems, components, and their interrelationships in order to better understand a given system. Some writers, however, obviously mean systems *synthesis* from their use of the term; others feel it relates only to computer systems, which is nonsense. Many use it in a sense that puzzles an engineer, since they seem to mean what he has known for a long time as the design process. When portraying systems analysis, its enthusiasts frequently speak of a "total system approach" to a problem (meaning a broad definition), of an intensive search for alternatives, of benefit-cost ratios; yet you recognize these as features of the process described over the last three chapters. So what's so new and different about systems analysis?

The term is ill-chosen, ambiguous, and carelessly bantered about, so frequently, in fact, that it could hardly be ignored here. Confusion is added by frequent use of the terms *systems engineering* and *systems approach* in a similar vein with similar disagreement among authors. Obviously these terms are more stylish than meaningful.

EXERCISES

1. Formulate and analyze the problem to which one of your existing courses is a solution. Preferably your choice will be the course for which this book is a text. This should precipitate a fruitful discussion of course objectives and of the backgrounds and attitudes of students at the start of this experience.

2. Choose some nontechnical problem other than those discussed in class or textbook and prepare the following: your formulation and your analysis. Some possibilities are problems associated with environmental improvement, health services, urban renewal, education, law enforcement, conservation, and athletics. Consider the pattern set by Figure 4-5.

3. You are a recent engineering graduate working in an engineering department. The chief engineer, realizing that you know your design technique well, and that some of the old timers around the department are sloppy in this respect, has assigned you to prepare a brief *Manual on Design*. Mainly in outline form, this manual will summarize the contents of Chapters 4 to 7.

4. From one of your present courses, choose a particular type of task (computational, prediction, measurement, writing, etc.) and define the problem at hand. Describe as many alternative methods of doing the job as you can.

FROM CLASSROOM INTO PRACTICE

Chapter 8

By now you have a rather rosy picture. You have learned design methodology, you conclude that it is a very logical and effective procedure, and you appreciate its widespread applicability. But the time has come to admit that there are flies in the ointment. The bad news is that execution of this design procedure is much more of a challenge in practice than it is in the classroom. Engineers encounter pressures, temptations, and pitfalls in their work that often lead them to stray from what they learned from text and teacher. Therefore, this chapter focuses on key differences between the relatively uncomplicated classroom and the sharply contrasting organizational world, so that when you graduate and begin applying your design skills, you can be better prepared to cope with major detractors.

The theory is that *you* will be better prepared once you are familiar with some of the discoveries that an ill-prepared student makes when he graduates. For instance, one thing he learns as he leaves his classroom (if not sooner) is that not everyone speaks the same design language. The situation can be frustrating. If you talk to an engineer and use terms like criterion and solution variable, you may discover, after some communication confusion, that he is unfamiliar with those terms or attaches different meanings to them. There are terminology differences from author to author and organization to organization. In fact, if you are using this book as a text for a course, don't be surprised to find that your instructor prefers terms different from mine (no reflection on him or on me).

Another discovery he makes when he arrives on the job is that not all engineering students are taught the same approach to problems. Although many educators and practitioners believe a systematic problem-solving approach will yield superior results in the long run, there is less than general agreement on the exact nature of that procedure. For instance, all authorities agree that you should define problems before you try to solve them. But most of them are frustratingly fuzzy about what constitutes a problem definition, and those who do offer specific guidelines differ in what they recommend. It's a fact of life—not everyone gets the same training—but like the non-standard terminology, it surely is not the most consequential of the student's postgraduation discoveries.

The big shock comes when it dawns on the student that solving problems on the job is by no means as straightforward as it is in the good old classroom. Since the difficulties gang up on him during his problem definition and decision-making efforts, it is worth dwelling on obstacles to effective execution of these phases of the design process.

Defining Problems In the Face of Numerous Obstacles

One reason defining real problems is not as straightforward as textbooks and instructors may make it seem is that those problems tend to be so effectively camouflaged—by irrelevancies, misleading opinions, past and present solutions, and unhelpful customary ways of viewing those problems. Then, too, problem definition on the job is hardly aided by the fact that problems in college are ordinarily presented in an unrealistically pure form, so that students are grossly unprepared for defining real-world problems. For instance, when problems are assigned, students almost invariably are handed exactly the information they need—no more, no less. Yet, in practice, much of the readily available information is irrelevant, even misleading, and much of what the engineer really needs (weightings of criteria, for instance) he has to almost pry out of the appropriate people. Also, there are present and past solutions to the problem at hand to cloud and mislead, plus those persistent troublemakers, fictitious restrictions.

Fictitious Restrictions

Consider this problem in which you are asked to connect these nine dots by not more than four straight lines without removing your pencil from the paper while drawing the lines. Some people cannot solve this problem, and others require a long time to do so, because they unjustifiably and perhaps unwittingly rule out the possibility of extending the lines beyond the square formed by the dots. They behave as if this were not permitted, even though no such restriction was mentioned in the statement of the problem. The unjustified, undesirable ruling out of a perfectly legitimate solution or class of solutions is a fictitious restriction.

Most fictitious restrictions are not explicit decisions to disregard certain possibilities. They are usually a matter of unconsciously assuming that certain possibilities have been ruled out. Many would proceed to solve the feed problem (page 64) *as if* the feed must be handled in sacks, even though no one said so. If such restrictions are explicitly stated, their fictitious and frequently absurd nature become obvious. Here are some characteristics of the current solution to the feed problem rephrased as restrictions: "Feed *must* be placed in sacks; the sacks *must* be handled individually." These are nothing more than characteristics of the presently used system with the words "must be" substituted for "is" or "are." The present solution to a problem is a likely source of imagined restrictions; the tendency to accept "what is" as "what must be" is strong, indeed. Probably most of us—even after we've been cautioned—fall prey to fictitious restrictions more often than we realize, or would admit if we knew.

Those Persistent SOLUTION Givers

Too often when an engineer is assigned to a problem it is a *solution* he is given and not a problem at all. This is a source of much difficulty. The following case illustrates this point *and* some good engineering.

City X is plagued by a parking problem. Forty percent of its commercial district is parking lots. This has prompted city officials to hire a consulting engineer to design a 600-car, multistory parking facility. It will be one of a series of similar facilities to be constructed in the congested area.

Note that the engineer has been given the city fathers' solution to the problem. His task is to detail this solution so that it is structurally and functionally sound. In spite of this, before specifying details of the desired facility, this engineer devotes some thought to the problem to which the proposed structure is *a* solution. He views the underlying problem as that of transferring a large segment of the population between its place of residence and its place of business. There is a major difference between this formulation of the *problem* and the restricted *solution* given to the consultant. His broad formulation opens up the problem to a whole realm of solutions. One is a high-speed transit system. Of course, nothing in this engineer's formulation precludes the possibility of a different type of urban community that reduces the need for mass transportation.

Under the circumstances he does what his "professional upbringing" dictates. He informs city officials that in his opinion increased parking capacity is not the obvious solution and that for the time being he is reluctant to design the proposed facilities. Instead, he outlines his view of the problem and some alternative solutions stemming from it. Adopting a broad view of the problem and standing by it when this course is in the best interest of the client is a mark of a professional.

Certainly not all engineers respond so beautifully under such circumstances. Some take it lying down and thus serve as technical detailers. Others offer only token resistance. Still others respond like the above engineer (and as a consequence, sometimes lose out on the job!).

But what's the point of paying a fat fee to a professional engineer and then proceeding to tell *him* what the problem is, which itself is rather presumptuous? Some, like the officials in this case, go a step further and tell him what the *solution* is.

Employers and Clients with Acute Restrictivitus

Recall the cleaning machine problem. It should have occurred to you that this engineer is operating under a fair number of restrictions set by previously made decisions. He is used to it; many of his problems come laden with restrictions. Some engineers welcome this; it reduces the number of decisions they must make. To others, it's a source of frustration.

Certainly not all imposed restrictions are accepted by engineers. This is far from true. Taking the cleaning machine problem of Chapter 4, it is possible to design a cleaning machine that will satisfactorily process *all* types of fabrics (and you can imagine how happy that would make marketing people!), but the expense involved in developing and manufacturing a truly general-purpose machine would be high, indeed. The engineer believes the probable selling price of a machine that processes all fabrics is disproportionately higher than the probable price of a machine that processes *almost all* fabrics. He is convinced that the "process-all-fabrics" decision is unsound. So he expressed this opinion, rather forcefully in fact, his objective being to have the original decision revoked. Sometimes he wins such cases, sometimes he loses. (Here he lost, but then, this cleaning machine never got off the ground, either.)

It is healthy for the organization and, moreover, it is a professional obligation to challenge restrictions that strike you as questionable, which might well constitute a majority of them.

An Organizational Fact of Life

In this book you read, "Formulate your problems broadly, it's your prerogative, your professional obligation to do so. Why shouldn't you? After all, it's a relatively brief mental exercise. Certainly there is no harm in *thinking* big." But a broad view of a problem in the mind is one thing; to what extent you are able to pursue it is quite a different matter. Pursuance of a broad formulation can bring you into direct conflict with decisions already made by your client or employer, or it can lead you into decision areas that are considered the responsibility of other persons in the organization. The engineer assigned to the feed distribution problem encountered resistance when he attempted to pursue his broad "producer-to-consumer" formulation. He had to persuade persons responsible for such decisions to give up the idea of individual sacks, to change the methods of warehousing, to alter sales policy, and the like. He succeeded but, for a variety of possible reasons, someone might well have told him to mind his own business, forcing him to pursue a narrower formulation that was not in the best interest of the enterprise. Many engineers are shot down when they attempt to fulfill this responsibility.

Now that you know something about real-world obstacles to effective problem definition, surely you will not underrate the difficulty of defining real problems or the importance of developing your problem-definition skill. Those who base their judgments of such matters on their classroom experience, where it probably seems simple and obvious, do underestimate the importance. It hasn't occurred to them that the classroom is a playpen in this respect.

Obstacles to Effective Decision Making

There are many, but one is prominent in my mind these days. Recall the young engineer guiding city officials in their selection of a solid-waste management system, and the laudable list of criteria he proposed to city officials. What do you suppose he ran into, recalling that some of the criteria called for a not-so-natural concern for downwind and downstream "nonusers"? Or can you imagine their response to criteria like expandability that imply a long-term concern for the city's welfare that officials elected to short terms are not renowned for displaying? And surely you are not surprised to learn that officials ignored his contention that they should give consideration to conservation of resources for distant generations.

The result was that he had to do some fast talking on behalf of criteria other than those arising from a short-term concern for effects within city limits. He failed to win his case, but did he fail professionally? There is no generally accepted answer to that one. Some would certainly say he went beyond the call of professional responsibility, many would be satisfied, and some would insist that he fell short of his professional obligation by not pushing hard enough. (Now you are getting a taste of the complexity of, and controversy surrounding, the matter of professionalism.)

The Bind

In classrooms, textbooks, and professional journals engineers are urged to consider the *total* cost of each alternative solution in decision making. Total cost requires that due consideration be given to unquantifiable as well as quantifiable criteria, indirect as well as direct effects, and consequences for nonusers as well as users over the long run as well as the short. Yet when engineers get on the job, they sometimes appear to have forgotten these classroom enjoinders. Have they?

Sometimes, yes, and for doing so they deserve the criticism they get. But more often they have probably had the same kind of experience the young engineer had with city officials. You can almost hear him saying: "Regardless of what you college professors say, or what I read in the journals, or what Ralph Nader thinks, if city officials have their minds *set* on a sanitary landfill, or if they are *not* concerned about conserving energy and material resources for future generations, if I disagree and it boils down to *their* way or *my* job, I could end up looking for a new job every six weeks if I didn't compromise in such binds. Once in a while an engineer learns who is boss." Similar words have been echoed a thousand times over.

In Sum

Nothing said in this chapter is cause for distress or discouragement. At the risk of same, these matters are mentioned so that you won't be as naive as I was at graduation. *You* want to be prepared for problem solving "under fire," not merely for handling classroom problems. There is a secondary motive: to let you know that the day-to-day work of the engineer is "livelier" than you heretofore realized.

Here is another good reason for not being distressed or discouraged. Although it is true that what prevails in practice is something less than what teacher and textbook prescriptions dictate, *don't for one minute conclude that engineers are any different in this respect*. Physicians, lawyers, and other professional people spend a lot of classroom time learning "how it should be done," only to find that beyond the classroom, they fall short. They do so for reasons somewhat different but no less compelling than those in engineering.

EXERCISES

1. Think about the prevailing manner in which problems are assigned to you in school and the bases on which your solutions are usually graded. Analyze what there is about those methods that leaves you unprepared for real problems.

2. Reflect on your own behavior, that of teachers and others around you, and society in general, and try to produce a list of at least 5 fictitious restrictions (or, for some cases, what amounts to same).

3. Seek out a puzzle like that cited on page 134 that provides a good example of a fictitious restriction at work. Write it up.

4. One portion of a large oil company warehouse is devoted to the storage, packing, and shipping of road maps to service stations. The current procedure is diagrammed in Figure 8-1. The stored maps are removed from their cases (200 to a case, 8 bundles of 25 maps each), and a moderate supply for each state is stacked on the open shelves. The packer, with order slip in hand, fills the customer's order by picking the requested type and quantity of maps from the shelves and assembling these in a carton on the bench. When the order is assembled, he slides the carton to the sealing and labeling station and performs these operations. Then he slides the carton to the next position in order to weigh the shipment and add the required postage. He then carries the completed order to the shipping dock. Under this

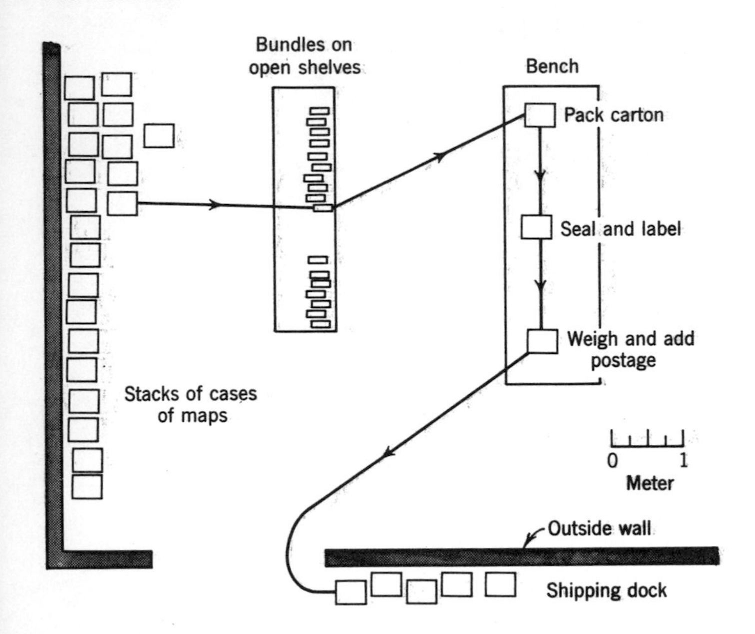

Figure 8-1

method, the typical order requires an average of 10 minutes for completion.

As the engineer assigned to improve this procedure, what do you recommend? (The process must be performed in the same part of the warehouse as that presently used.) Adequately describe the workroom layout, the procedure, and the equipment that you propose. *Also,* provide an outline of the manner in which you arrived at your solution, including your problem formulation and analysis, and the alternatives you considered. This is one of those cases in which you are not being handed all the information you need. Your instructor will provide the facts you need, assuming that you ask the appropriate questions.

MODELS

Chapter 9

The following objects have something in common: toy train, global replica of the earth, statue, molecular model, and toy airplane. Each is a three-dimensional *representation* of a physical reality. There is a two-dimensional equivalent, as exemplified by photographs, sketches, and blueprints. Since these two- and three-dimensional representations bear a physical resemblance to their real-life counterparts, they are referred to as *iconic representations*. Engineers make frequent use of iconic representations; you can see why the one pictured in Figure 9-2 is a valuable aid. Can you picture it: trying to visualize what this structure looks like from 189 pages of complicated engineering drawings?

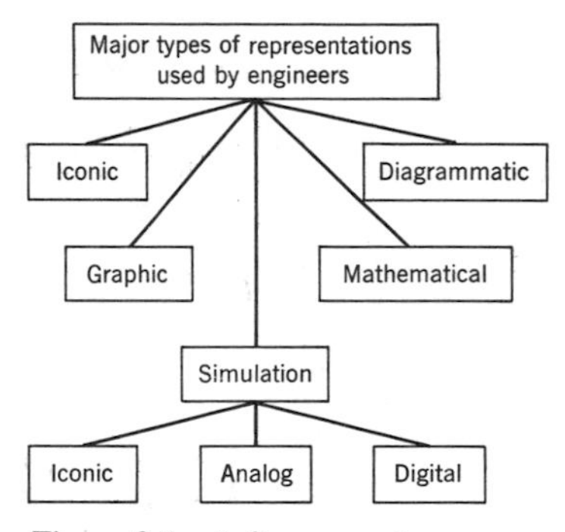

Figure 9-1 *A diagrammatic representation of the contents of this chapter.*

Then there is the familiar *graphic representation* (Figures 9-15 and 9-18). You are already familiar with the usefulness of graphs in aiding you to visualize relationships and relative magnitudes.

The diagram for a football play, the schematic diagram for an electrical circuit, and Figure 9-1 are *diagrammatic representations;* so is Figure 9-22. In each instance a configuration of lines and symbols represents the structure or behavior of a real-life counterpart. A diagram such as Figure 9-3 is certainly helpful to those planning and managing the project, especially because of the number of interrelationships. As you can imagine, with engineering devices, structures, and processes becoming so complex, diagrams must be relied on extensively in designing these systems and in communicating their make-up to others.

Mathematical Representations

$$V = \frac{mkT}{p}$$

The mathematical expression shown here is a representation. The letter m represents the mass of a quantity of gas, T represents its temperature, p represents the pressure applied, V represents the volume occupied by that gas, and k is a constant. Together these letters represent what happens to one of the properties when a

Figure 9-2 *An iconic representation of an oxygen-production facility. This table-top model aids the designers in laying out piping especially so that things will be accessible for maintenance and repair. It is also used as a visualization aid to them and to persons to whom they must explain their design.*

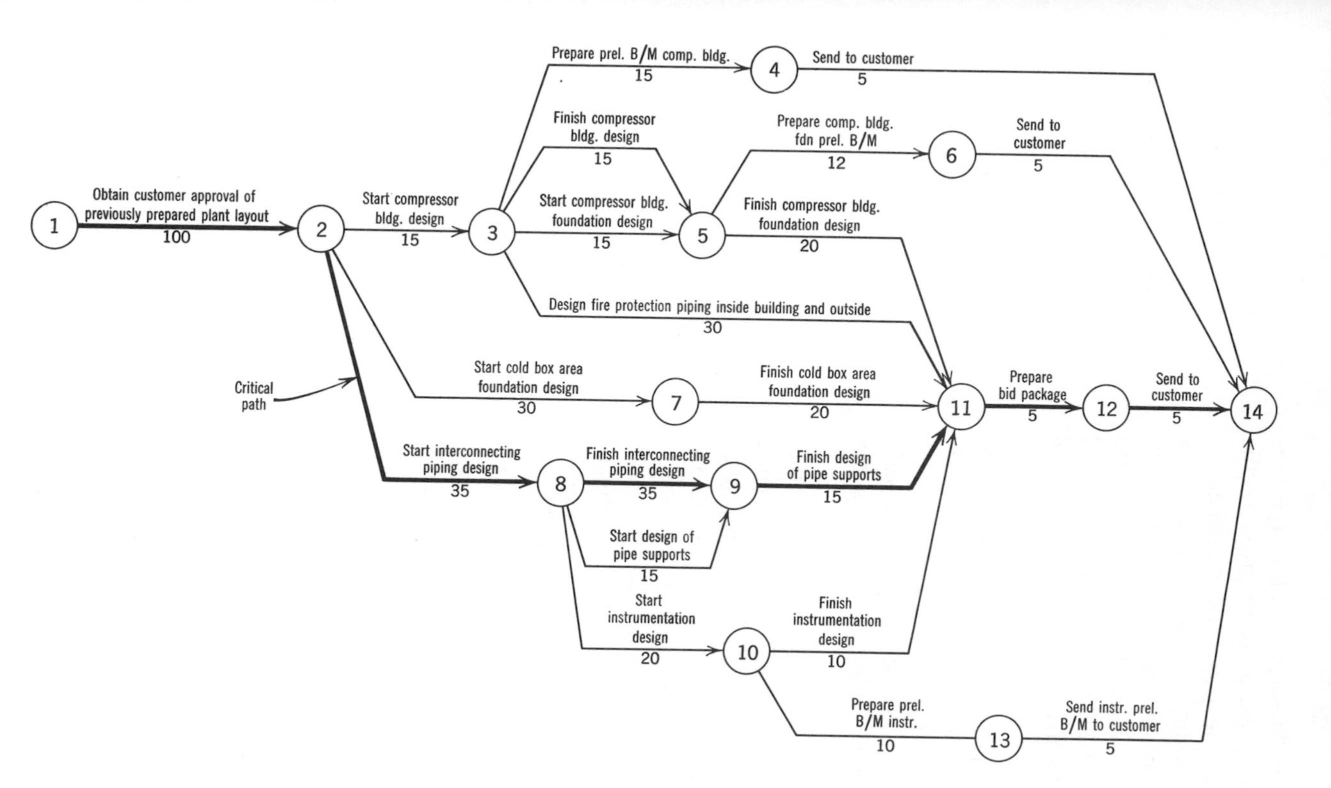

Figure 9-3 *A diagrammatic representation of a portion of an engineering project. Arrows leading to a circle represent those phases of the project that must be completed before activities represented by arrows leading away from that circle can begin. The number beneath an arrow is the number of days that phase of the project is expected to require for completion. There is one "path" through this network, referred to as the critical path, which determines the minimum period of time in which the total project can be completed. This type of representation, commonly referred to as a critical path diagram, is a very useful aid in planning and controlling large-scale design and construction projects. The project diagrammed above involves design of the foundation, piping system, and building for a chemical plant. As critical path diagrams go, this is simple. But even this one demonstrates the value of diagrammatic representations as aids to visualization of relationships.*

specified change occurs in another. This *mathematical representation* provides a means of predicting one property, given specific values of the others. Through the power of mathematics, predictions can be made of many natural phenomena as well as of the behavior of man-made mechanisms and structures. By employing the conventions of mathematics and by assigning symbols to represent relevant properties of the real thing, mathematical expressions can be developed for making useful predictions of what to expect under given conditions.

The study of mathematics will provide you with a repertory of ready-made mathematical representations (parabolic function, sine function, etc.). It also equips you to derive special mathematical expressions to fit situations that cannot be represented satisfactorily by ready-made mathematical functions. This is a very important skill for engineers.

Mathematics provides a powerful means of representation, serving as an effective means of *prediction* and as a concise, commonly understood language for *communication*. Its conventions also make it an extremely useful *medium for reasoning*. Can you imagine trying to perform in words some of the logic and the manipulations that you can do so conveniently through the symbolism of mathematics? In addition, training in mathematics strengthens your ability to think clearly and logically. In view of the utility of mathematics as a means of prediction, communication, and reasoning, the heavy emphasis given to this subject in engineering education is readily understandable.

Simulation

An iconic representation can be used to predict the behavior of its real counterpart. A model of a proposed aircraft (Figure 1-14) is subjected to high-velocity winds in the wind tunnel in order to predict how a real plane of that design will perform in actual flight. What the wind tunnel and plane model are to the aircraft designer, the facilities pictured in Figure 9-4 and 9-5 are to the designers of bridges, tall buildings, power plants, and ocean-going vessels. The same is true for engineers planning flood control measures (Figure 9-6). The process of *experimenting with a representation* of the real thing is called *simulation*. When

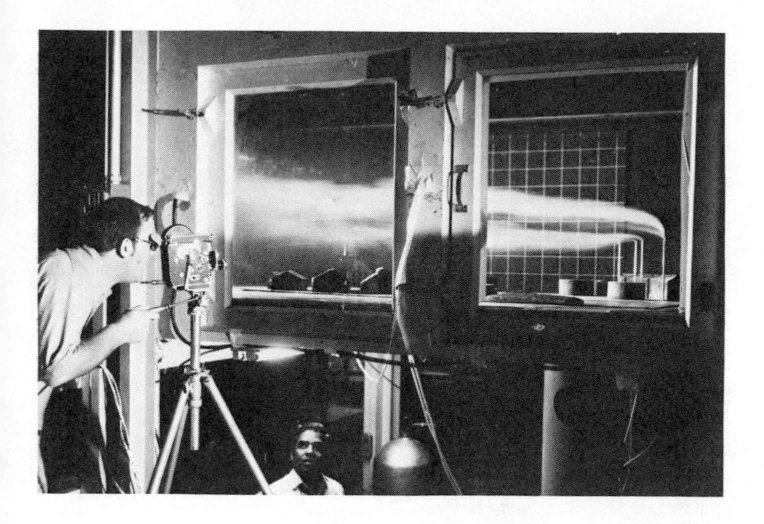

Figure 9-4 *Wind tunnels are useful for more than testing aircraft. Here, through the viewing window of a wind tunnel, you see a scale model of a power plant under design. Engineers are experimenting with different stack heights to determine what is necessary to disperse the stack emissions satisfactorily. Also tested in wind tunnels are bridge designs for wind-induced oscillations, models of tall buildings to predict the wind currents they will create, and models of ocean liners for optimum location of stacks.*

Figure 9-5 *In this basin, models of ocean-going vessels are tested for maneuverability and seaworthiness. The carriage that tows and maneuvers the models rides along a 115-meter bridge that itself is movable. Special machines generate waves of specified size and frequency. This simulation setup enables engineers to predict the full-scale performance of proposed ship designs on the high seas.*

Figure 9-6 *This is one section of an enormous working scale model–the largest in existence, representing the Mississippi River and its tributaries from Sioux City, Iowa, to the Gulf of Mexico. This simulator is used to predict the local and system-wide effects of proposed dams, diversionary channels, and other construction projects.*

the experiments are performed on iconic representations, the process is called *iconic simulation*.

There are other forms of simulation in which the experiments are performed on representations that bear a behavioral but not a physical resemblance to their real-life counterparts. One of these is called *analog simulation,* another is called *digital simulation*.

An example of analog simulation is the electronic device used by an engineer who is designing a traffic control system. Special electrical circuits represent the city's traffic arteries, while electrical pulses represent vehicles. With this simulator he experiments with different traffic control schemes. Here electrical pulses behave analogously to autos moving about the city, even though pulses and wires in no way physically resemble autos and streets. In the analog simulator pictured in Figure 9-7, water behaves analogously to air. It enables designers of gas turbines to test their ideas quickly and cheaply. Thus, in analog simulation, a medium that behaves analogously to the real phenomenon is employed as a vehicle for experimentation. Electricity is the medium often used.

Digital simulation is best introduced by an example. A university plagued by auto parking problems has engaged a consulting engineer to improve the situation. One of his proposals is to separate the drivers who consistently use their parking spaces the full working day from those who use their spaces sporadically (e.g., a few hours one morning, all the next afternoon, an hour the following morning). The full-day users will be assigned regular *spaces,* but the sporadic users will be grouped and assigned to a *lot* where they will use any available space. The engineer theorizes that more of the sporadic drivers can be assigned to a lot than there are spaces, with negligible risk of the lot overflowing. Notice, however, that this is an untested idea. How does he go about verifying his hypothesis?

One way is to set up a special lot, assign only sporadic users to it, and then observe it for a significant period of time to see what happens. This is too costly, too involved, and too time-

Figure 9-7 *An analog simulator. In this instance water behaves analogously to air and serves as the medium for experimentation. The device simulates air flow through the diffuser blades (represented by wedge-shaped objects) in the compressor stage of a gas turbine under development. Water containing a dye that makes the path of flow readily observable is diffused from the center at one thousandth of the velocity of the gas it represents. By experimenting with different wedge shapes, angles, and locations, the investigators will learn how to maximize the effectiveness of this part of the engine.*

TABULATION SHEET FOR PARKING SIMULATION — Day No. 1

Identification of Row	Source of Table Entries	Driver No. 1	2	3	4	5	6	7	21	22	23	24	25
Show or no show?	Spinner												
T_a (arrival time)	Spinner	8.5	10.0	8.9									
P_e (expected park time)	$P_e = 10.2 - .56\,T_a$ (or use graph)	5.4	4.6										
D (chance deviation)	Chip drawn from bowl	+.5	−1.1										
P (actual park time)	$P = P_e + D$	5.9	3.5										
T_d (departure time)	$T_d = T_a + P$	14.4 (2:24)	13.5 (1:30)										

Figure 9-8 *Here is the crux of the digital simulation. In executing this simulation the operator performs the indicated numerical operations, using this sheet as a running record of the experience he has synthesized. (Hours are in decimal form, so that 12 midnight = 0, 9:30 a.m. = 9.5, and 3:00 p.m. = 15.0.)*

consuming. A logical alternative to this real-life experiment is to test the theory by simulation, which the engineer did. In this simulation he assumed a 20-space lot to which 25 drivers were assigned. His procedure and results are described in Figures 9-8 to 9-11. (The details of digital simulation are explained in the Appendix to this chapter.)

Thus, by simulation, the engineer was able to evaluate the effects of a policy of overassignment—before making any *real* changes—*and* he had something to back up his proposal when presenting it to his client. It took a clerk two days to perform this simulation, employing a simple pencil-and-paper procedure; gathering data and setting up that procedure required another three days. Thus, in five days, it was possible to synthesize 100 days of experience, illustrating the "time compression" capability of simulation.

Figure 9-9 *This sequence of steps diagrammed at the right is repeated 25 times to complete the simulation of one day of parking-lot operation. Then another day is simulated, and another, until sufficient experience in the operation of this hypothetical parking lot has been synthesized. The results for one simulated day are shown in Figure 9-10. The spinner device is used to recreate the element of chance associated with events. The use of this device and the drawing-from-the-bowl procedure are ways of making random selections from a collection of numbers, appropriately referred to as Monte Carlo Procedures. (See Appendix to Chapter 9.)*

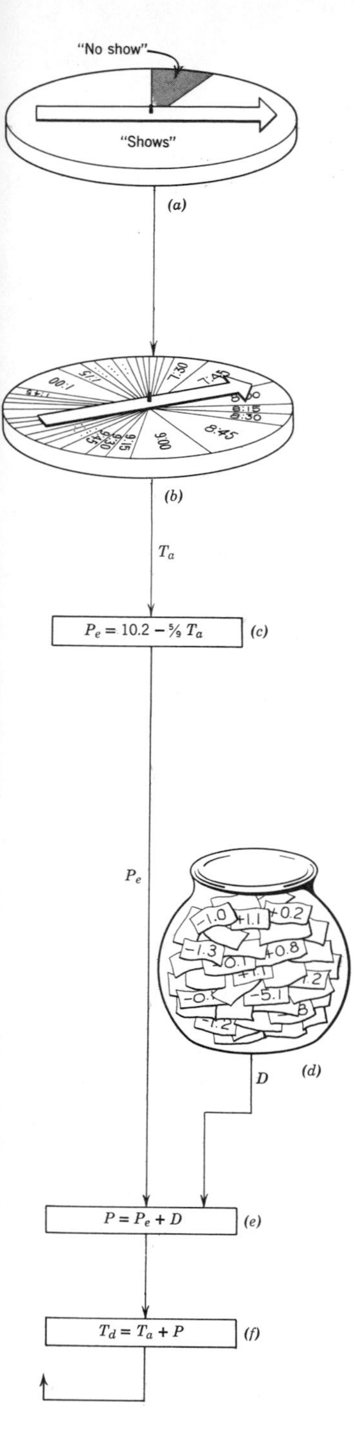

(a) From a survey the engineer estimates the probability that a driver will not use the lot on any given day as 8 chances in 100. This "no show" probability of 0.08 is taken into account by using this roulette-type device. Beneath the pointer is a pie chart having a "no show" sector of 28.8 degrees (0.08×360 degrees). This pointer is spun for each of the 25 drivers, to determine whether he uses the lot on that day.

(b) The likelihood of a driver arriving is not the same throughout the day. To determine how arrival frequency changes with time of day some actual observations were made. These data were converted to this pie chart in which each sector is proportional to the average arrival frequency of drivers for the indicated time of day. This pointer is spun to select an arrival time (T_a) for each driver.

(c) The length of time an auto remains on the lot depends on the time it arrives. To learn the nature of this relationship some "park times" were measured resulting in the graph below. A straight line was fitted to this data. From the

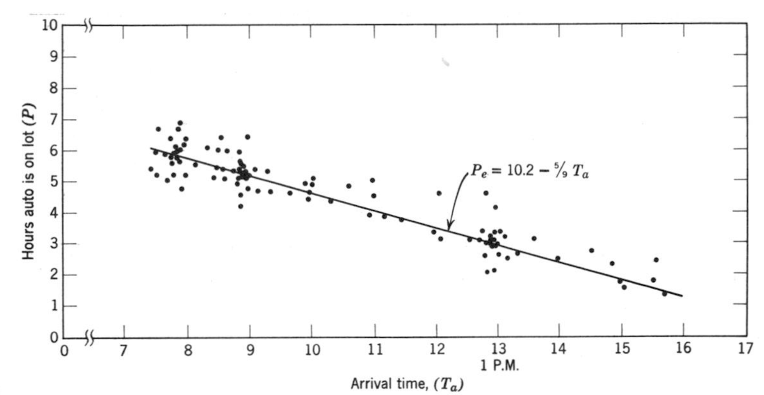

equation of this line the *expected* (most likely) park time (P_e) can be computed, given the driver's time of arrival. (P_e is the average parking time for many drivers arriving at a given time of day.)

(d) From the above graph it is apparent that the *actual* periods of time autos remain on the lot vary considerably from the *expected* times (i.e., from the straight line). The vertical deviation of each point from the straight line was measured (plus if above the line, minus if below it). Each of these deviation values was recorded on a separate slip of paper and placed in a bowl. In the simulation a slip is selected randomly, the D value noted, and the slip returned to the bowl.

(e) Then this D value is added to the *expected* park time (P_e) to obtain the *actual* park time (P).

(f) The departure time (T_d) of a given driver is calculated by adding his park time to his time of arrival.

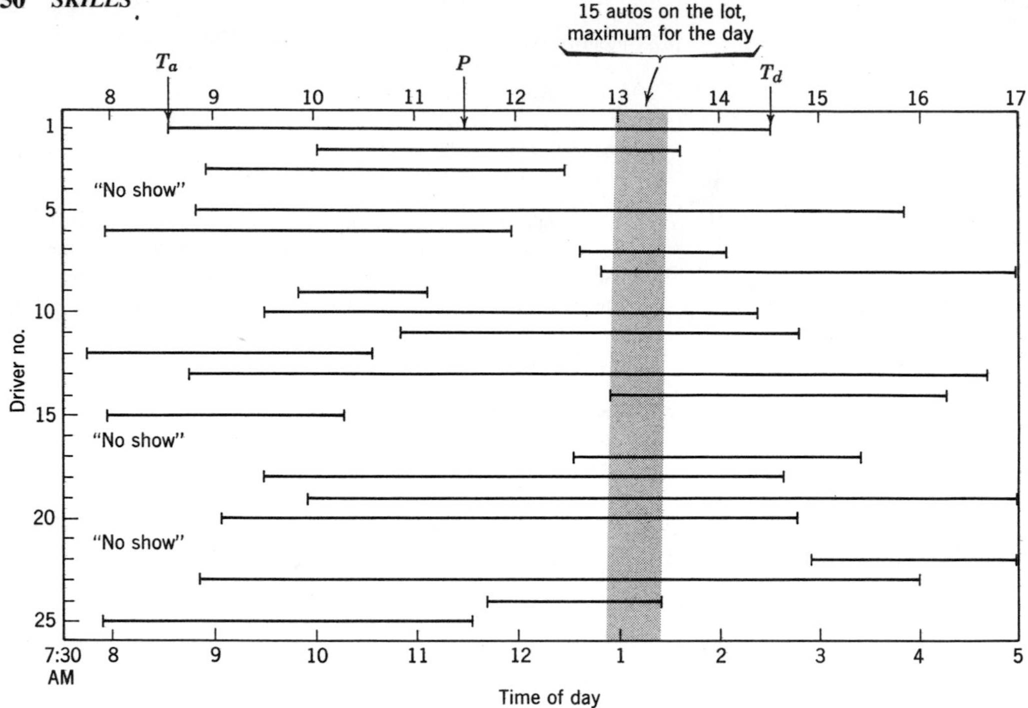

Figure 9-10 *A graphic representation of one day of simulated parking-lot operation. The maximum number of autos is determined by noting the maximum number of simultaneously overlapping bars.*

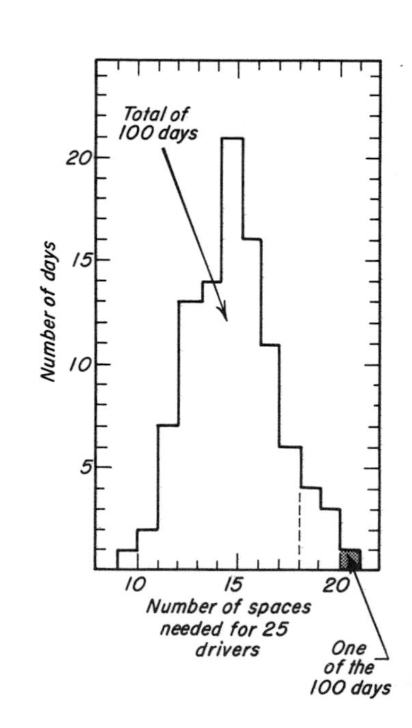

Figure 9-11 *This graph is a summary of 100 daily maximums. Each daily maximum is determined as explained in Figure 9-10. Note that on only one of the 100 days simulated did the lot overflow, even though 25 drivers were assigned to a 20-space lot. Furthermore, on only 4 percent of the days did the daily maximums exceed 19 autos, and on only 8 percent of the days would the lot have overflowed if there had been 18 spaces. From this it is apparent that more drivers of the sporadic type can be assigned than there are spaces on the lot, resulting in better utilization with negligible inconvenience to the users. Furthermore, the client can decide what probability of overflow he will tolerate; then the engineer can use this graph to determine how many drivers can be assigned to a given lot without exceeding this probability. Suppose university officials say that they don't mind if the lot overflows 25 percent of the time. For this not to be exceeded, 16 spaces should be provided for a group of 25 drivers of the sporadic type. (Don't let this 25 percent alarm you; even when the lot overflows, only one or two drivers are usually affected, and there are other lots for them to go to.)*

Digital simulation is experimentation with a digital representation, an "acting-out process" in terms of numbers. It is simple yet remarkably powerful. Because it is a series of step-by-step numerical operations, it can be executed by a computer. This is fortunate. Doing this by pencil and paper is laborious and time-consuming. Actually, digital simulation would be prohibitively expensive in many potential applications if computers were not available to perform these highly repetitive numerical manipulations. Simulation by computer has become popular in engineering, as illustrated by this sample of uses: simulation of the behavior of atomic particles, of space flights, of aircraft movements around an airport, of rush-hour operations of a battery of 95 high-speed elevators in a new skyscraper, of all types of traffic phenomena, and of global warfare.

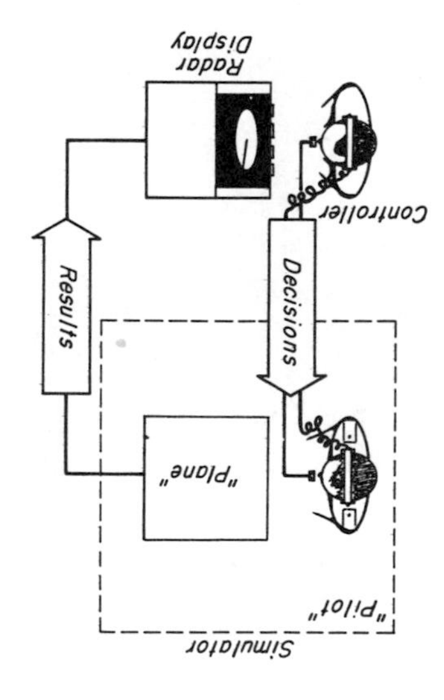

Figure 9-12 *Pilots and planes? Well, not really, but as far as the traffic controllers in the next room are concerned, they are. Each person in this room is playing the part of a pilot, and the special apparatus in front of her is a "plane." As she "flies" her flight plan (e.g., destination, course, altitude) by manipulating the controls on the console, this information is sent by wire to a controller's radar set, where it causes a pip to move realistically across his screen. The controller-pilot conversations take place over telephone lines between the two rooms.*

Participative Simulation

It is possible to involve humans directly in a simulation. Picture a group of air traffic controllers in a room. In front of each controller is a radar screen on which the usual pips appear, representing the positions, identities, and altitudes of planes in his area of control. Each controller is in radio communication with these planes, issuing instructions to and receiving reports from planes in his zone of responsibility. To the casual observer, it all looks like the real thing: a busy traffic control center. Yet the "pilots"

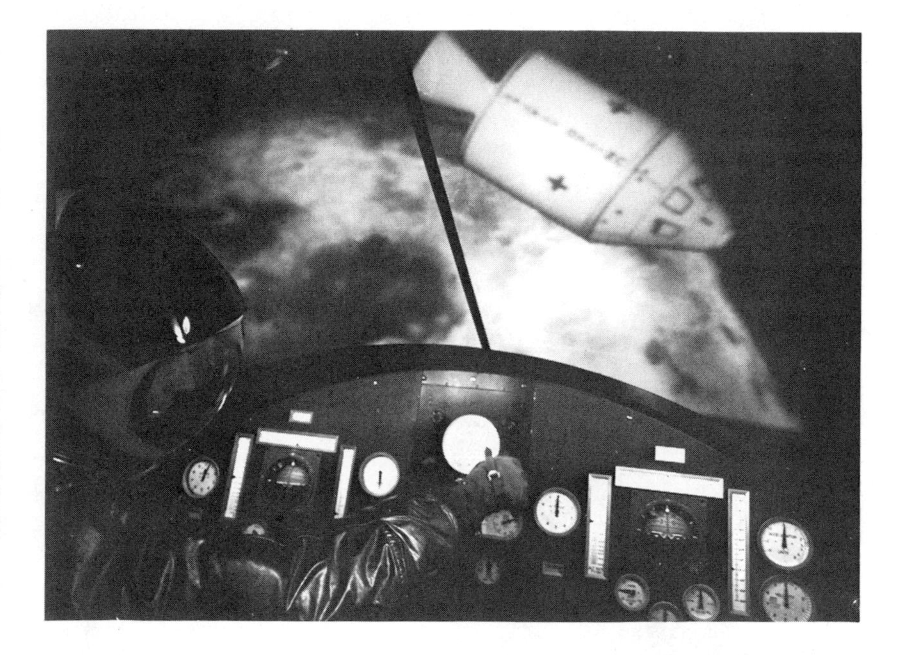

Figure 9-13a *This astronaut, shown maneuvering his spacecraft during rendezvous with another vehicle, can go home at 5 P M with other employees of this aerospace company because he is performing this mission on a simulator in the laboratory. The moon and spacecraft images visible to him are being projected onto a large, semicircular screen in front of his "vehicle." As the astronaut manipulates his controls, the scene in front of him changes appropriately, providing him with the illusion of movement with respect to the moon and to the other vehicle. In an adjacent room a TV camera, focused on a three-dimensional model of the moon, picks up the image that appears on this large screen. The camera moves in response to the astronaut's manipulation of the controls. A similar closed-circuit TV system projects the vehicle on the screen. Launches, landings, and orbital missions can be simulated in order to learn what astronauts can and cannot do and to train them for the real thing.*

with whom these controllers are communicating are in the next room (Figure 9-12).

These "pilots" and "planes" constitute a large simulator used by the team of traffic controllers in the next room to test new traffic control procedures. The whole thing operates like this. On the basis of the traffic situation in the controller's zone of jurisdiction, as it appears on his radar screen, he makes decisions and communicates them to the appropriate "pilots." The resulting maneuvers of their planes become apparent on the controller's screen. The controller continues with his decisions and instructions, and the pilots continue to respond, creating a situation that is very realistic, at least for the controllers.

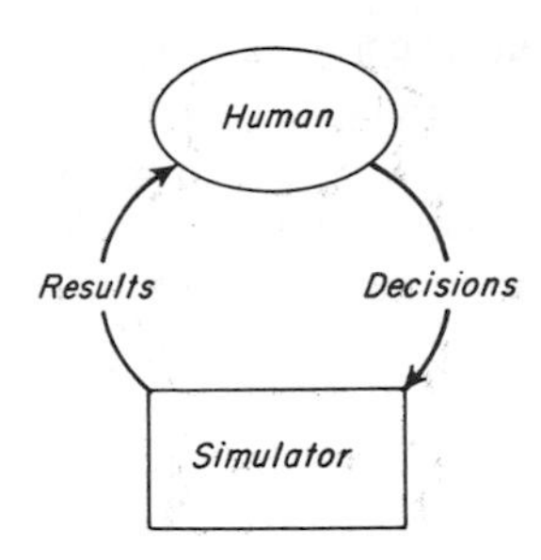

Many simulations directly involve people as decision makers, following the same pattern: the human participant (e.g., a pilot in

Figure 9-13b *If you were to come across one of these while taking a stroll, you might look twice. This is a model of a modern supertanker employed to train pilots in the handling of such vessels. The "bridge" has all the essential instruments, and realistic time lags are built into the controls (it is mainly these long ship-response times that make piloting of large ships such a challenge). Water facilities at this site enable trainees to practice docking and other maneuvers under normal and emergency situations. A wave-making machine creates the equivalent of 6-meter waves.*

a trainer) communicates decisions (the pilot through his controls) to the simulator, which determines and communicates the results of those decisions through visual displays, as a continuous cycle. Many driver trainers work on this principle. Figures 9-13 and 9-14 provide other examples.

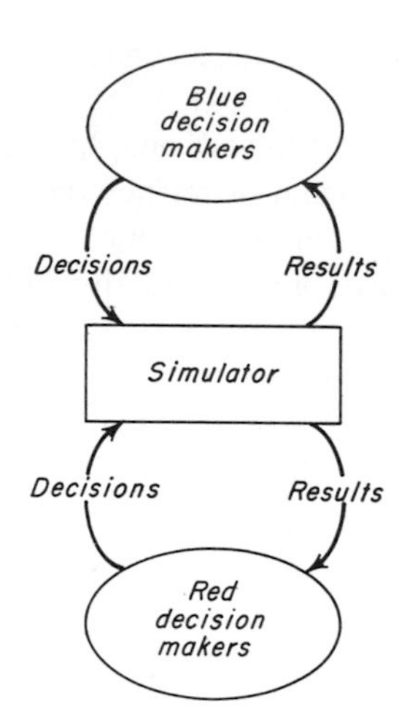

Since a human being participates directly, this is termed *participative simulation*. If two or more persons compete with one another in a participative simulation, it is usually referred to as *gaming*. Participative simulation is useful for prediction and for training purposes. (Most of us would prefer to have a fledgling pilot make his learning mistakes on a simulator!)

Thus, an engineer can experiment on iconic, analog, or digital representations in order to make predictions about their real-life counterparts. This process, simulation, is a means of synthesizing experience by operating a representation for a period of time to learn how the real thing will perform. This often beats experimenting with the real thing in several respects; it costs less, takes less time, enables engineers to exert closer control over their experiments, and is less hazardous (imagine how rapidly the aircraft industry would consume test pilots if it didn't rely heavily on simulation).

Models

The representations described here are usually referred to in the literature and conversation of engineers and scientists as *models*. So we ordinarily speak of graphic models (instead of graphic representations), diagrammatic models, and so forth. This requires that you extend your interpretation of the word model beyond ordinary usage. To the engineer and scientist, a *model* is anything used to describe the *structure* or *behavior* of a real-life counterpart. Models do this through words, numbers, special symbols, diagrams, graphs, or by looking like or behaving like the real-life counterparts that they represent.

It may take awhile for you to appreciate fully the generality of this concept. When it does dawn on you, it will be clear that all of the following are models in the sense that this term is used by scientists and engineers: a mental image of a person or of an experience or of anything; our conception of the nature of light; Darwin's theory of evolution, Einstein's relativity theory, *any* theory; the words dog, stick, book, stone, and most other definite nouns; a verbal description of the workings of a mechanism; a musical score; and a chemical formula. Then you will not be

Figure 9-14*a* *An airport runway at night? Well, not exactly.*

Figure 9-14*b* *The scene in Figure 9-14*a *was taken in this building at night from the "cockpit" shown. The cab moves up and down, left and right, and lengthwise in response to the "pilot's" manipulations of the controls. The runway lights visible in Figure 9-14*a *are located on the floor of this room. This 300-meter-long facility is used to simulate landings in fog in order to predict the effectiveness of lighting and other guidance systems. A spray system creates a fog of desired density.*

puzzled if you hear someone refer to Darwin's model, the wave model of light, or your model of engineering.

The term model is used universally in engineering and physical science; it is also becoming common in the language of social and biological scientists. So, for communication with persons in these fields, it helps to know what they mean when they talk about the so-and-so model, since they will invariably assume that you are accustomed to using the term model in a broad sense. But the contents of this and the next chapters are important for more than terminology reasons. The ability to employ models for the purposes described in the upcoming chapter is a powerful skill that you can almost certainly put to use, regardless of your occupational specialty.

In Summary

Many models—mathematical equations, three-dimensional models, graphs and diagrams—are certainly familiar enough to you. In these instances what may be new to you is the point of view: that these tools have something in common as expressed in the term *model*. However, the situation might well be different with digital simulation; here you may have learned a new technique, an elegant one, to be sure; it is amazingly useful yet surprisingly simple. Of course, digital simulation by computer can't be old hat—computers haven't been around that long. So if you know something about digital simulation, and especially if you know how to prepare such simulations for computer execution, you are probably one up on most of the elders in whatever field you go into.

MONTE CARLO METHODS

Appendix

A barber would like to know, *before* he purchases the equipment and hires the new man, what effects adding a second barber to his shop will have on customer waiting and barber idleness. Digital simulation is an effective means of making the predictions he desires. To set up a digital simulation for him I proceeded as follows.

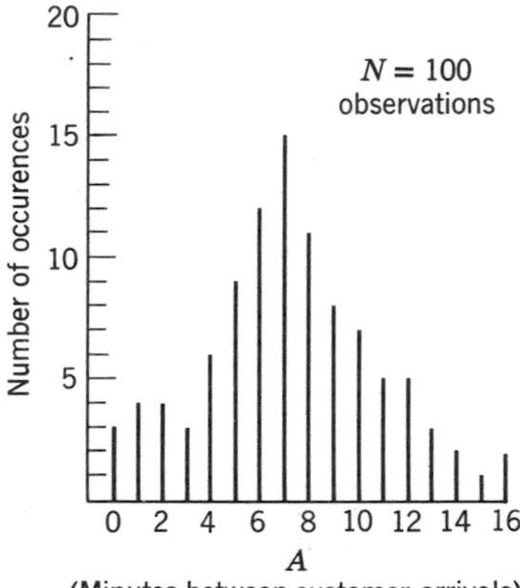

Figure 9-15 *A frequency histogram summarizing the 100 observations made of minutes between successive customer arrivals at the barber shop.*

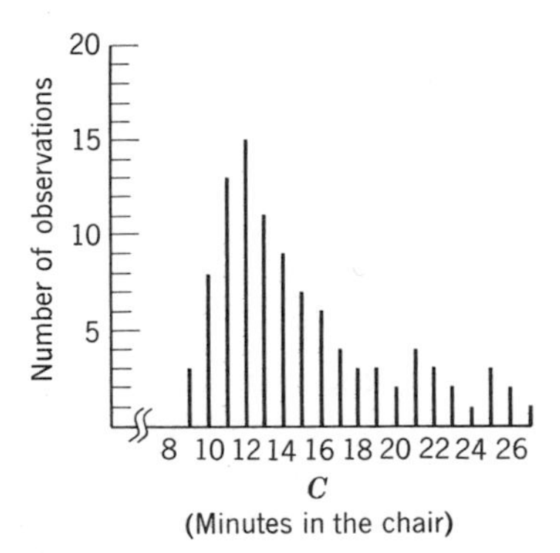

Figure 9-16 *A frequency histogram summarizing the sample of "chair times."*

1. **I observed the present one-barber operation to collect a sufficient sample of times *between* successive customer arrivals and of haircut times, observing long enough to record such times for 100 customers. The resulting *A* values (times between arrivals) and *C* values (chair times) are summarized by Figures 9-15 and 9-16.**
2. **Then I took 100 slips of paper and wrote the 100 *A* observations on them, which meant there were three slips with zero on them, four slips with 1 on them, four slips with 2 on them, and so forth.**
3. **I did likewise for *C* times.**
4. **I set up a table (Figure 9-17) to provide a chronological record of events about to occur as this shop is operated on paper by a process called digital simulation.**
5. **The simulation model was set in motion. To do so, I had before me the two piles of slips and the tabulation sheet. I assumed for this test run of the model that there was one barber and that the shop opens at 8 a.m. Literally, I picked an *A* value at random from that pile, and used it to determine the time of the first customer's arrival. The value selected (and returned to the pile upon reading it) was $A = 10$ minutes, and so the time of his arrival was**

$$T_a = 8{:}00 + 10 = 8{:}10$$

That fact was so recorded on the tabulation sheet. Of course, since he was the first arrival, there was no

TABULATION SHEET

	Customer number →	1	2	3	4	5	6
A	Value selected	10	15	4	7	13	15
T_a	Arrival time	8:10 (8:00+10)	8:25 (8:10+15)	8:29 (8:25+4)	8:36	8:49	9:04
T_c	Cutting starts	8:10	8:25	8:37 (previous T_e)	8:55		
C	Value selected	14	12	18			
T_e	Cutting ends	8:24 (8:10+14)	8:37	8:55			
W	Customer wait	0	0	8 (8:37−8:29)	19		
I	Barber idle	10 (8:10-8:00)	1	0	0		

Figure 9-17 *Tabulation Sheet for barber shop digital simulation.*

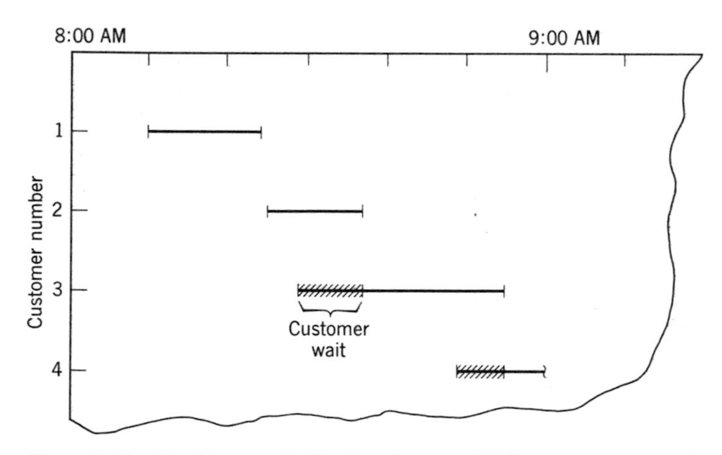

Figure 9-18 *Graphic record of barber shop simulation.*

waiting, and so $T_a = T_c$, where T_c is the time at which the first customer gets in the chair.

6. **In similar fashion, a C value of 14 minutes was selected and recorded in the table and, from that, the time at which the barber was finished with customer one was determined as**

$$T_e = T_c + C = 8{:}10 + 14 = 8{:}24$$

These times can be found on the graph of Figure 9-18.

7. That procedure was continued until enough experience had been synthesized to justify conclusions. Notice from Figures 9-17 and 9-18 that customer waiting occurred occasionally. At the end of the experimental run, these wait times were summed, and so were barber idle times.
8. But hold on. There are methods of making these random selections of A and C—called *Monte Carlo methods*—that are more sophisticated and convenient than the slips-of-paper routine. One of them involves a pie chart and free-spinning pointer. One was constructed for generating A values by replotting Figure 9-15 in circular form. The resulting pie chart (Figure 9-19) is equipped with a free-spinning pointer and becomes a perfectly acceptable method of choosing values randomly from a population of numbers.
9. But there is a better way, based on a table of random numbers. Picture yourself with a spinner device like that shown in Figure 9-20, which has equally spaced marks numbered 1–100 around the circumference. You spin the pointer and record its stopping point in terms of this 1–100 scale. The results of many repetitions of this procedure are shown in Table 9-1.

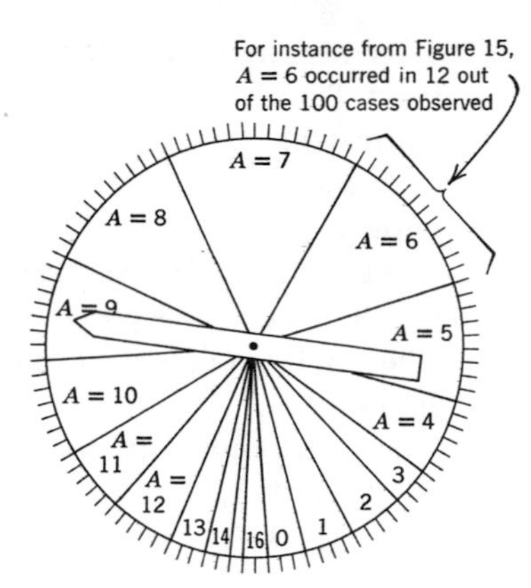

Figure 9-19 *Pie chart equipped with free-spinning pointer in order to select randomly "time until next arrival."*

TABLE 9-1 TABLE OF RANDOM NUMBERS

12	82	89	54	11
74	41	21	02	13
81	06	19	79	91
19	43	17	75	82
29	21	35	18	57
47	98	81	96	28
26	60	24	77	49

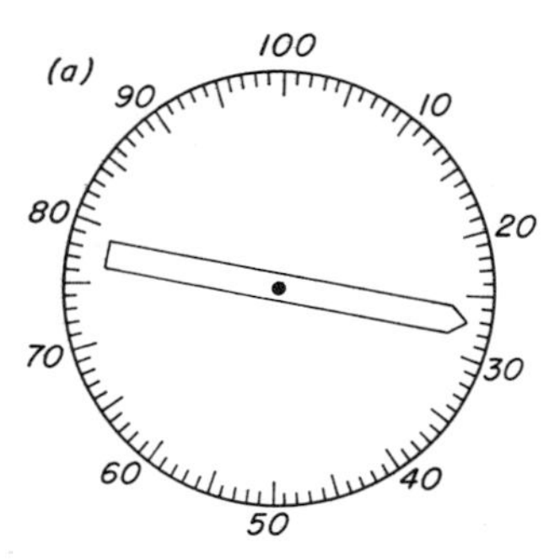

Figure 9-20

In *the long run,* each of the numbers 1–100 will appear with equal frequency in a table of random numbers, but in random order. Such tables are readily available in textbooks and handbooks.

This table can replace the pointer pictured by Figure 9-19. By numbering from 1 through 100 around the circumference of the pie chart, creating Figure 9-21, it is now simply a matter of selecting a number (you can go across rows, down columns, or pick at random, it matters not) from Table 9-1 and consulting Figure 9-21 to

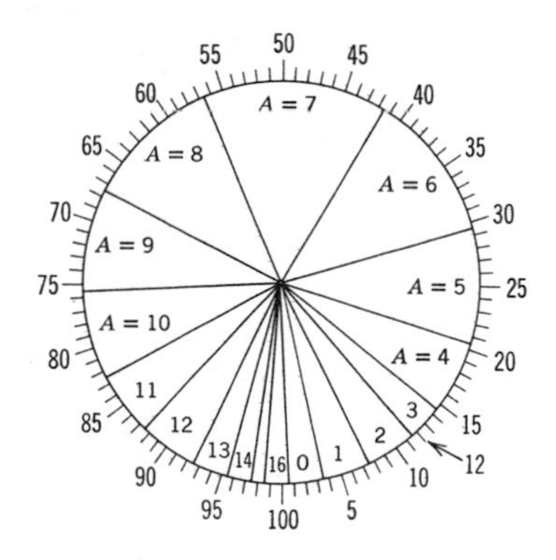

Figure 9-21 *The pie chart of Figure 9-19 with the spinner replaced by a circumferential scale of 1–100.*

determine in what sector that number falls. If I select 12 from Table 9-1, that falls in the $A = 3$ sector. This is equivalent to the pointer of Figure 9-19 stopping in the $A = 3$ sector *and* to selecting a 3 from the 100 slips of paper.

10. But I can do still better. For the procedure described in step 9, not even the pie chart of Figure 9-21 is necessary. The whole process can be reduced to a series of rules, as follows.
    Select a random number *(RN)* from Table 9-1. Then:
    IF RN is 1–3, $A = 0$ minutes
    IF RN is 4–7, $A = 1$ minute
    IF RN is 8–11, $A = 2$ minutes
    IF RN is 12–14, $A = 3$ minutes
    .
    .
    .
    IF RN is 99 or 100, $A = 16$ minutes
    These rules can be conveniently summarized into what is commonly called a *conversion table* (Table 9-2).

TABLE 9-2 CONVERSION TABLE

IF random number *(RN)* is:	*THEN* time until next arrival *(A)* is:
1–3	0 minute
4–7	1 minute
8–11	2 minutes

Thus, with a table of random numbers (1 through 100 in this case) and a conversion table, I have the most convenient of the "pencil-and-paper" Monte Carlo methods. This method, the slips of paper routine, and the pie chart and spinner mechanism are equivalent.

11. A digital computer can also make random selections from a population of numbers. One way a computer can do this is diagrammed in Figure 9-22. In this case, random numbers are supplied externally; they are on punched cards and available whenever the computer "asks" for one. The procedure diagrammed here can be explained in terms of the pie chart of Figure 9-21. The computer asks for a random number and then, in effect, works its way around the circumference, starting

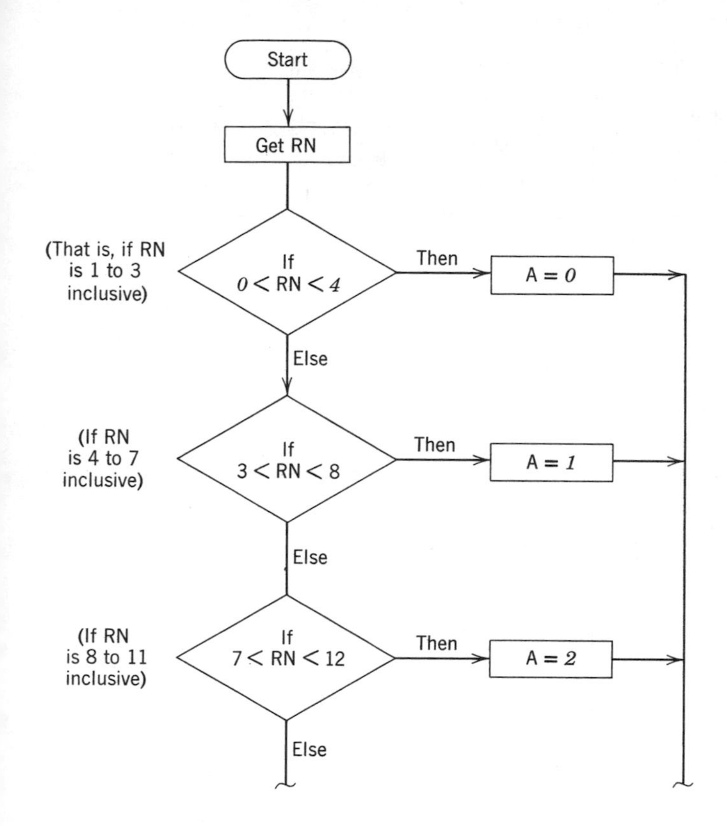

Figure 9-22 *Diagram of a portion of a computer routine for Monte Carlo-ing. If so desired, the computer can generate random numbers internally through a subroutine.*

with the $A = 0$ sector. It asks: "*RN*, do you fall in this sector?" If the answer is yes, it assigns $A = 0$, but if the answer is no, it goes to the $A = 1$ sector. If the answer there is yes, A becomes 1 minute; otherwise the machine goes to the next sector. It continues around the circumference, stepping from sector to sector until it scores a "hit" and assigns the appropriate A value.

If you understand the procedure just described, and if you are able to program a computer, you are equipped to write computer simulation programs. This is no small matter, for simulation by digital computer is a very powerful technique of rapidly increasing importance.

EXERCISES

1. A machine for twisting strands of wire into a cable must be shut down rather frequently because of breaks and jams. When it does, an attendant must repair the fault before the machine can get back into production. An engineer is setting up a simulator to determine how many attendants should be assigned to a battery of nine such machines. Set up a Monte Carlo device to generate breaks (actually times between breaks) and a different Monte Carlo mechanism to generate repair times. Data is supplied by Tables 9-3 and 9-4. Explain how the engineer might use these random generators in his model of this battery of machines.

2. The management of an oil company is contemplating a modification of one of its tanker unloading facilities at a cost of several million dollars (Figure 9-23). However, some of the

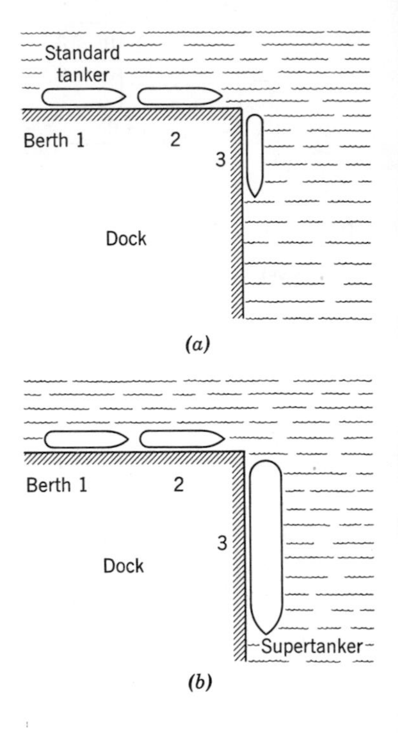

Figure 9-23 (a) *At present the company can accommodate three standard tankers at its unloading facilities.* (b) *The proposed modification will enable the company to unload supertankers at Berth 3.*

TABLE 9-3 DATA ON INTERVALS BETWEEN WIRE BREAKS

Minutes Between Breaks	Observed Frequency
5	7
10	19
15	28
20	21
25	13
30	9
35	3

TABLE 9-4 DATA ON BREAK REPAIR TIMES

Repair Time (minutes)	Observed Frequency
6	12
7	32
8	29
9	17
10	10

consequences of this change are not evident. Therefore they have hired you to answer questions like:

(a) *What effect will the elimination of one "standard berth" have on the average time tankers spend waiting for unloading?*

(b) *Should standard tankers be assigned to the "superberth" when it is the only place available to unload at the time?*

This would be a trivial problem if tankers arrived at predictable times and unload times were uniform. But this is not the case (Figures 9-24 and 9-25), so predicting the effects of the proposed changes is not that simple.

Here, as in most such instances, you have at least four methods of predicting the performance of the facility proposed: judgment, building and trying it, mathematics, and simulation. The first will not be accepted *in this case* primarily because the investment is so high. The second alternative is absurd *in this case.* The third is infeasible *in this case* primarily because you would have to develop the equations (which would prove to be prohibitively complex here). So you are left with simulation, digital simulation, in fact.

One of your first steps in setting up a digital simulation model would be to collect data on behavior of the existing system. This has been done for you, with the results shown in Figures 9-24, 9-25, and 9-26. You may safely assume that this data applies to the system proposed. You have been told that 1/6 of the arriving tankers will be "supers." That is all the data you need to set up and operate a digital simulation of the proposed system.

To generate a significant amount of experience would require that you operate your model for longer than you want to spend. Therefore operate your model only long enough to demonstrate to yourself and your instructor that you know what you are doing, say for 10 days of synthesized experience. Do so for each of these experiments.

(a) *A facility with two standard berths and one superberth (Figure 9-23*b*), under a policy that permits standard tankers to unload at the superberth when the two standard berths are already occupied.*

(b) *The same facility as in part* a, *but under a policy that requires standard tankers to unload at standard berths.*

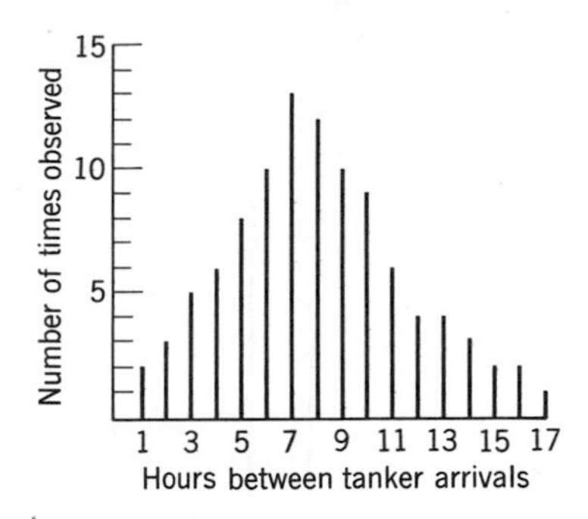

Figure 9-24 *A graphic model of a sample of times between tanker arrivals. These values were obtained directly from 100 successive tanker arrivals as recorded in the dockmaster's log. This, therefore, is past experience in tanker arrivals, which you assume will hold under the proposed setup.*

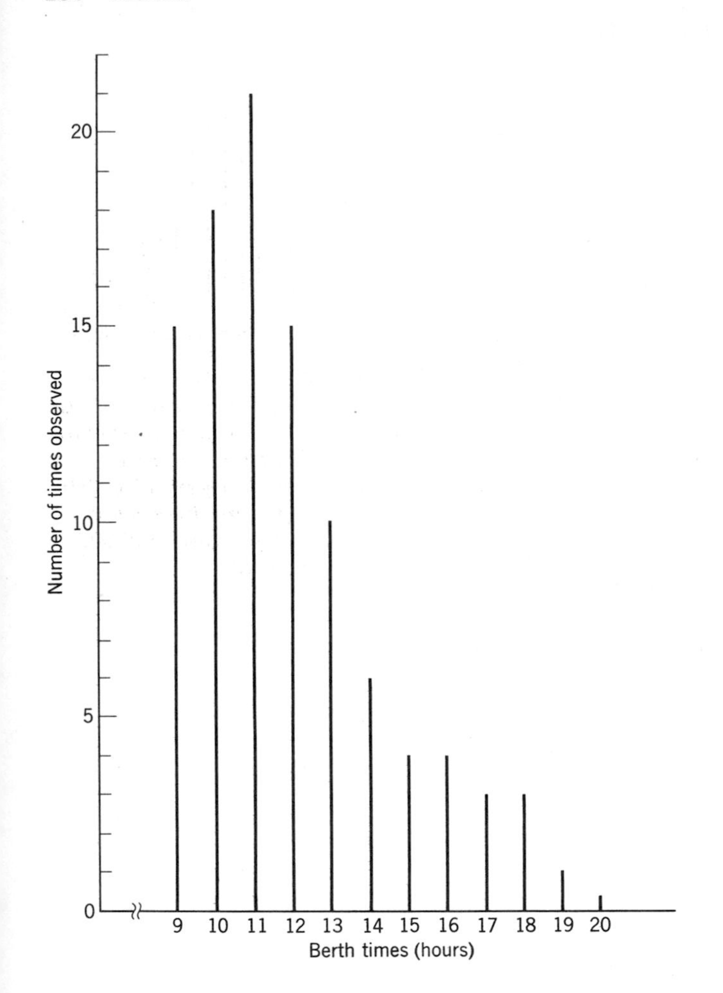

Figure 9-25 *A graphic summary of berth (unload) times for 100 standard tankers, obtained from the dockmaster's log.*

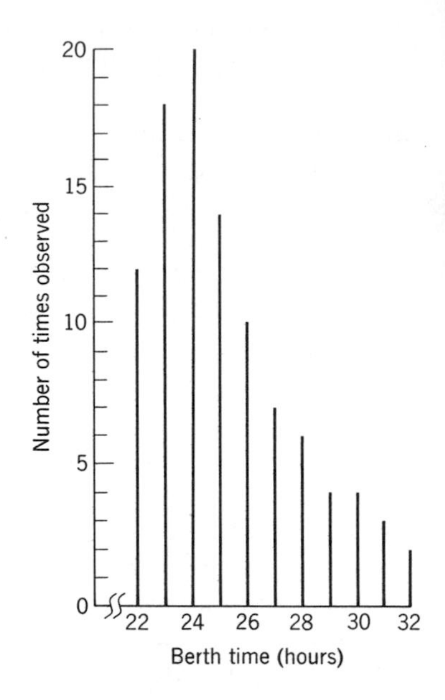

Figure 9-26 *A graphic summary of berth times for 100 supertankers, obtained at another facility this company operates for unloading supertankers. (An important assumption, which seems reasonable since the pumping equipment will be identical, is that this sample satisfactorily characterizes unloading times at the proposed facility.)*

After running these two experiments, total the hours tankers were forced to wait for unloading. Under the assumption that the company loses $200 for every hour a standard tanker is idle and $700 for every hour a supertanker is forced to wait, calculate the total cost of waiting for policy *a* and for policy *b*. To repeat: these figures are not statistically significant; 10 days of "experience" is not enough for practical decision-making purposes in this case. Want some help? I assume you do. It's available in Figures 9-27, 9-28, and 9-29.

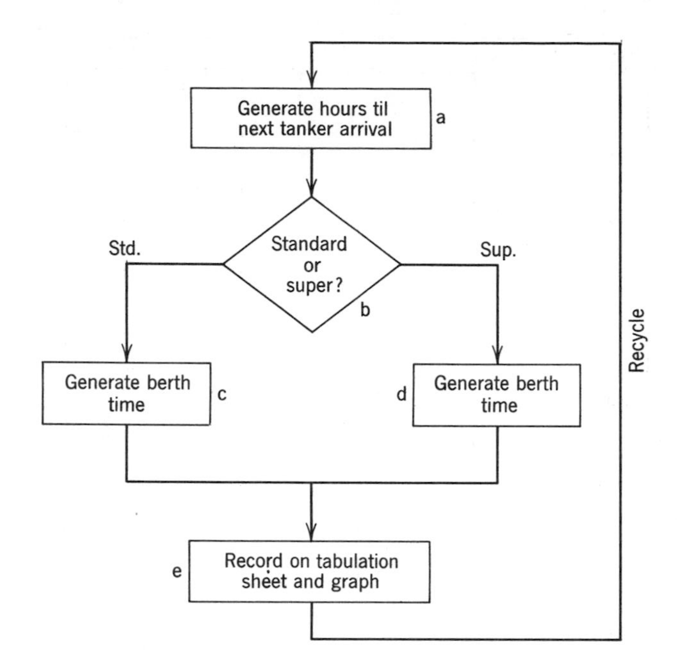

Figure 9-27 *A way of setting up the tanker simulation. Steps a, c, and d require random selection of values from the data supplied by Figures 9-24, 9-25, and 9-26. Try your hand at each of the Monte Carlo methods introduced in the Appendix to this chapter. Which one you use where is up to you. The tabulation sheet and the graph referred to at point e are alternative means of keeping a running record of events; try your hand at both. Suggestions for these are made in Figures 9-28 and 9-29.*

3. Survey other textbooks that you are using (or have used) and identify three examples of each of the following types of models: iconic, graphic, diagrammatic, and mathematical. A brief verbal description of the illustration will suffice. Identify the book and the page on which each example is found.

4. Monte Carlo methods—in words, what are they designed to do?

5. The number of yards gained was recorded for the last 20 times football play X was used by the Purple Tigers. The *average* (expected) gain for this sample of 20 is 5 yards. But chance plays a role; there is considerable chance variation about that 5-yard average, which is indicated by this frequency histogram (Figure 9-30). In simulating a football game, chance can be introduced, and to do so for this play we will want to select deviations

	Tabulation sheet												
(a)	Time between arrivals	2	4	15	4								
	Arrival time	2	6	21	25								
(b)	Tanker type	Su	St	Su	St								
	Start berth 1 (Standard tankers only)		6										
	Start berth 2 (Standard tankers only)				25								
	Start berth 3 (Supertankers only)	2		28									
(c) or (d)	Berth time	26	11	25	12								
	Berth 1 Ready		17										
	Berth 2 Ready				37								
	Berth 3 Ready	28		53									
	Wait time	0	0	7	0								

Identification of experiment: Two standard berths, one superberth. "Standards" NOT allowed at superberths.

Figure 9-28 *Digital simulation tabulation sheet for tanker unloading operations. The lettered rows key them to Figure 9-27. I have generated the first four tanker arrivals, using Monte Carlo methods to generate times between arrivals and berth times, and a die to make the "standard versus super" choice. The event times for the first four arrivals are plotted in Figure 9-29.*

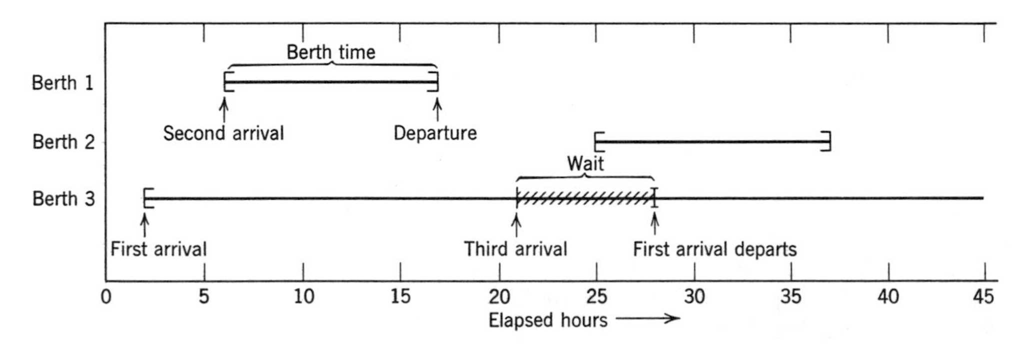

Figure 9-29 *A graphic record of the arrival, berth, and departure times for the first four columns of Figure 9-28.*

randomly from the values summarized by Figure 9-30. Set up and test run three different but equivalent Monte Carlo methods of making the desired random selections.

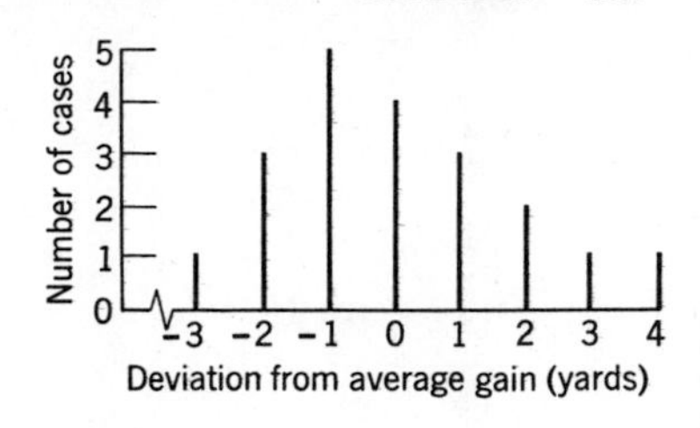

Figure 9-30 *Assuming a 5-yard average, these are the deviations from that average in 20 trials of play* X.

MODELING

Chapter 10

Models have great utility in engineering, so much so that the following overview of engineers' uses of models tells you much about the profession. In fact, one of the engineer's more conspicuous attributes is his quickness to turn to a model. It seems so natural to turn to a diagram to describe something, to a graph to make a point, and to mathematical symbols to aid thinking.

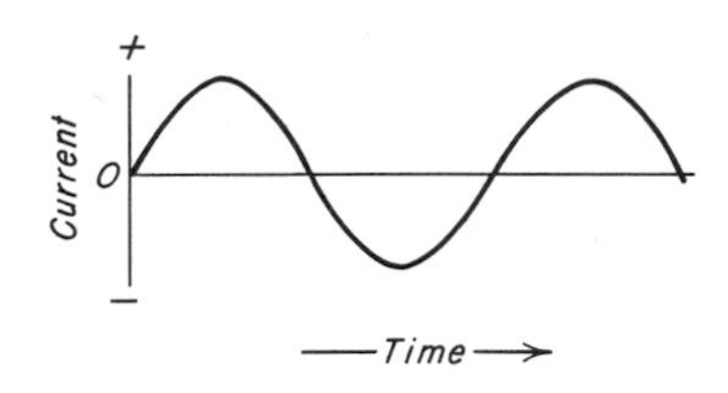

How Engineers Use Models

For Thinking

A model can be a real help when you are trying to visualize the nature or behavior of a system or phenomenon that the unaided mind finds difficult to grasp. There are electrical circuits, manufacturing systems, chemical processes, and mechanisms, the complexity of which makes a diagrammatic or other type of model essential to human comprehension. Iconic, diagrammatic, and graphic models are especially helpful in providing a compact, overall, simplified view of the whole. Often, in thinking of a physical phenomenon, the engineer finds it expedient and profitable to think of it in terms of a model. For instance, an experienced engineer usually thinks of an alternating electrical current as a sine wave in graphic form instead of as the movement of electrons in a conductor. You might think of a gust of wind in the vague terms of a chilling sensation on your face; but an engineer who designs aircraft or long-span bridges will most likely think of it in graphic form. Often it is these abstractions that engineers manipulate in their thinking. *A major objective of an engineer's education is to develop the ability to think of physical phenomena in terms of useful abstractions.*

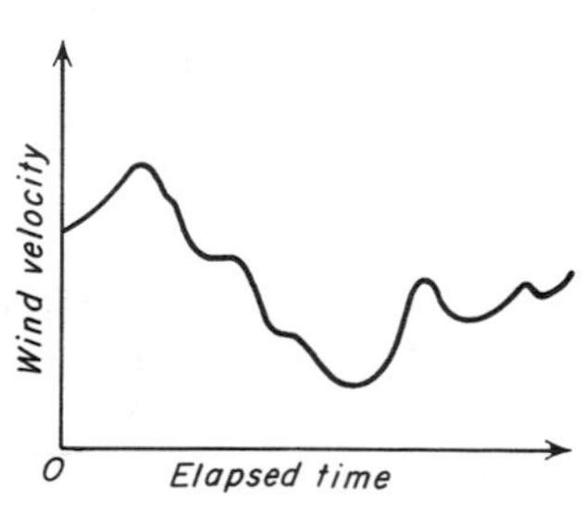

For Communication

Of course, most communication is through models. An author communicates to you through symbols, photographs, sketches, and diagrams. In addition to these common communication aids, engineers frequently use mathematics, graphs, three-dimensional models, and drawings. They rely on models especially when the systems and phenomena to be communicated are complex. Imagine trying to describe in words the system diagrammed in Figure 9-3.

For Prediction

Recall that a major part of decision making is predicting the performance of alternative solutions with respect to the criteria selected. Models are utilized extensively for this purpose; in fact, engineers would be lost without these fundamentally important tools.

Here is an example of the usefulness of predictive models in design. Several of the television transmission antennas atop the Empire State Building are to be replaced (Figure 10-1). The engineer in charge thinks it desirable to predict how much the modified antenna stack will bend as a result of wind load. There is no mathematical model specifically for predicting bending (deflection) of such antennas, but he (an electrical engineer) remembers something about the analysis of beams from basic engineering courses. He recalls that mathematical equations for predicting just about anything you would want to know about a beam rigidly fixed at one end—called a cantilever beam—are readily available in textbooks. He can treat the antenna stack as a vertical cantilever beam and use this model, which predicts end deflection *(d)* of a beam uniformly loaded over its length *(L)*.

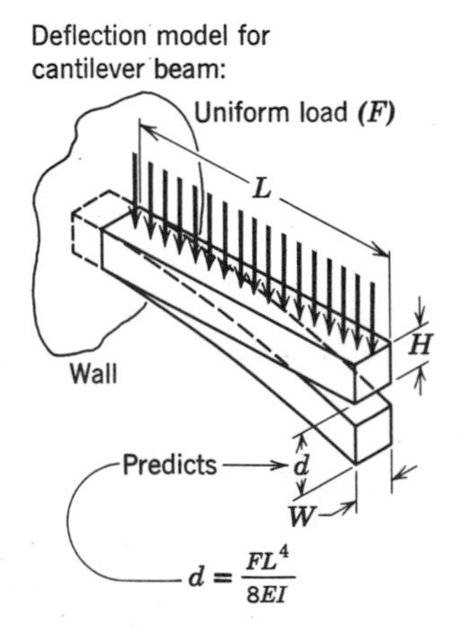

$$d = \frac{FL^4}{8EI}$$

$$d = \frac{280 \times 50^4}{8(2.07 \times 10^{11})(7.866 \times 10^{-3})}$$

$= .134$ centimeters

KRICK ... Concepts
CH.10, W. 152

Therefore, all that remains to predict d is to obtain certain data (e.g., normal and maximum wind velocities at this altitude), to make certain assumptions (e.g., wind resistance of the stack), and then to substitute these values into the deflection equation. You are being spared the details; they are unimportant at the moment. What *is* important is that you appreciate the utility of this model in *predicting* behavior of the structure while it exists only on paper; herein lie the beauty and the power of a predictive model.

Recall the Severn Bridge pictured on page 42. During the design of that bridge, extensive wind tunnel tests were made on alternative cross-sectional shapes (20 of them, in fact) of the bridge deck to arrive at a shape that minimizes wind effects. (And here is an interesting sidelight. Something that you and I might overlook is the effect of winds during the delicate operation of lifting and positioning the deck sections. Engineers simulated this operation in a wind tunnel and learned that special provisions had to be made to control wind effects during these operations.)

The above cases are typical in this respect: in solving problems an engineer must evaluate most solutions while they are still in the conceptual stage. Models are extremely useful for this purpose; they enable him to make the required predictions of solution performance without the necessity of physically creating the solution. Through the manipulation of mathematical and simulation models, it is possible to evaluate solutions with less time, cost, and risk than experimentation with the real thing ordinarily requires, yet more accurately than is usually attainable when pure judgment is used. For the engineer designing the antenna stack, he is not about to have a full-scale version of his design mounted atop the building just to measure deflection! Yet the stakes are too high to rely solely on opinion. Predictive models are an excellent compromise in such situations.

Figure 10-1 *Eight television stations have their transmission antennas stacked atop the Empire State Building. Several of these, near the top, are to be replaced with improved designs. That is a challenging problem.*

For Control

When developing a model for prediction purposes, an attempt is made to have the model's predictions agree as closely as feasible with what eventually occurs. In some situations the converse is true; an attempt is made to get the modeled situation to conform to the model. The engineering drawings for a building constitute a model and, of course, the building is constructed to conform to that model. The flight path a spacecraft must follow to reach its objective is carefully computed beforehand. This planned flight path is a model; very elaborate systems are employed to hold the spacecraft's actual flight path to that model.

For Training

Most models that are useful for communication are useful for instruction. Not so obvious, however, is the usefulness and growing popularity of participative simulation as a training tool, especially when the costs of blunders are high because of danger to life or the expensive equipment. Thus simulation is relied on extensively to train commercial and military pilots, air traffic controllers, and astronauts. The astronauts and all key ground personnel repeat their feat many times through simulation in the laboratory before the real countdown and blastoff. Similarly, the crews that operate our missile defense and attack systems get no opportunity to practice at the real thing, yet they must become proficient at their tasks. The answer is simulation.

An engineer's ability to employ a variety of models for the above purposes is important, indeed, since a wide variety of models serve a multitude of purposes in design. Observe some: the usefulness of diagrams in structuring problems (page 67), the value of sketches in creative thought (page 95), the role of mathematics and simulation in predicting the performance of alternative solutions (page 148 and 162), and the usefulness of most types of models for specification of the final solution.

Uses of Models in Other Fields

It is worth observing, as you probably did for the design process, the applicability of modeling skills to nonengineering problems. An important purpose of citing nontechnical extrapolations of skills such as design and modeling is to enable you to visualize specifically what people are talking about when they speak of the broad applicability of an engineering education.

You know that models are used extensively in science. A book in any branch of physical science, physics, for example, is packed with various kinds of models. This is also true for the biological sciences in which iconic models are conspicuous, and for the social sciences, where verbal models predominate. So this is no news to you. But one aspect of modeling that you do not encounter routinely in undergraduate, nontechnical textbooks is applications of digital simulation.

It is possible to simulate digitally: the behavior of neurons and neuron networks; cancer cell multiplication; movements and collisions of atomic particles; learning processes; consumer behavior; business activities of numerous kinds; economic activity at the regional and national level; meteorological phenomena; and global warfare. In fact, these are not only possible, they have been accomplished, in each instance with the help of a digital computer. So digital simulation does have widespread applicability outside of engineering.

Digital simulation by computer may subsequently serve a more important role in the biological and social sciences than it has in physical science and engineering. Chance is a significant factor in the behavior of living things, and the Monte Carlo technique is a means of synthesizing the role of chance. So individual and group behavior of all kinds of living organisms can be realistically simulated by a digital computer. This is especially significant because biologists and social scientists find it difficult to conduct satisfactorily controlled experiments when dealing with the actual phenomenon. Experiments performed on digital models can be controlled, as the following case will illustrate.

Suppose you had a computerized simulation model of a typical city, which you could employ to predict employment levels, housing demands, population shifts, changes in the local economy, and the like. Suppose further that you could experiment with this model to predict the consequences of various courses of action open to public officials, such as construction of low-cost housing, inauguration of job training programs, expansion of public transportation facilities, and alteration of the tax structure, *before* public officials take such actions. Furthermore, assume that these predictions are over the long run—say over the next 50 years.

If you had such a model you would have something of value to society. It would enable you to satisfactorily answer a multitude of "what if" questions that officials today must answer on the basis of intuition.* But surely, considering the complexity and consequences of such decisions, this is a lot to expect of intuition. Certainly experimenting with the urban system itself is an impractical means of determining the best course of action. Time is against it, and so are the number of simultaneous, uncontrollable changes that inevitably confound such experiments.

A model gets around these obstacles by permitting controlled experiments on a compressed time scale. Of course, you say,

* "What if question" is a popular expression used to describe the capabilities most predictive models provide: the ability to predict *what* will happen *if* a hypothesized change in action or policy is instituted.

"But no such model exists." Maybe not, but we are coming close. Professor Jay Forrester, for example, has reported a model that, in scope, approximates the model postulated.* Forrester's computerized model warrants attention considering the potential of his approach and in view of what he calls the counter-intuitive results of his experiments. Forrester reports:

> Various changes in policies were examined with the model to show their effect on an urban area. A number of presently popular proposals were tested—a job training program, job creation by bussing to suburban industries or by the government as employer of last resort, financial subsidies to the city, and low-cost housing programs. These all were shown to lie between neutral and detrimental in their effect on a depressed urban area. . . . As you can see, the results are controversial. If they are right, it shows that most of the traditional steps taken to alleviate the conditions of our cities may actually be making matters worse.†

Since the urban model employed by Forrester for these experiments is still relatively crude, do not conclude from his rather unsettling remarks that all urban programs presently in force should be scuttled. But this modeling approach shows enough promise to warrant continued refinement, with the expectation that it will evolve into a practical aid to public decision makers. Of course, such models will not provide *the* answers to problems of public concern but, when combined with imagination, new insights, and sound judgment, they will enable decision makers to do significantly better.

This urban model illustrates a point made earlier: that models have widespread applicability to nontechnical problems, including those of a social nature. However, public planners and decision makers have been slow to take advantage of such tools. Social systems—nations, corporations, schools, hospitals, cities, whatever—are extremely complex. And worse, authorities seem to be losing ground in their ability to forecast trends and to predict the probable outcomes of alternative courses of action for such systems. Our greatest hope for acquiring better predictions appears to lie in models, particularly in nontraditional models like computer simulation.

* See Jay W. Forrester, *Urban Dynamics*, MIT Press, 1969.

† *The Engineer and the City,* proceedings of a 1969 symposium by that title, sponsored by the National Academy of Engineering.

Model versus Real World

Some realities associated with the operation of a campus parking lot are ignored—"assumed away," as the saying goes—by the simulation model described earlier. As far as the model is concerned, weather has no effect on the probability of a driver using his car, yet we know this is not true in real life. Furthermore, there are some features of the parking lot model that are not true of the real thing. For example, it was possible in the model to generate a negative park time (it was ignored if it occurred), which obviously doesn't occur in real life. Such discrepancies between the model and its real-world counterpart are inevitable. They are found in *every* model.

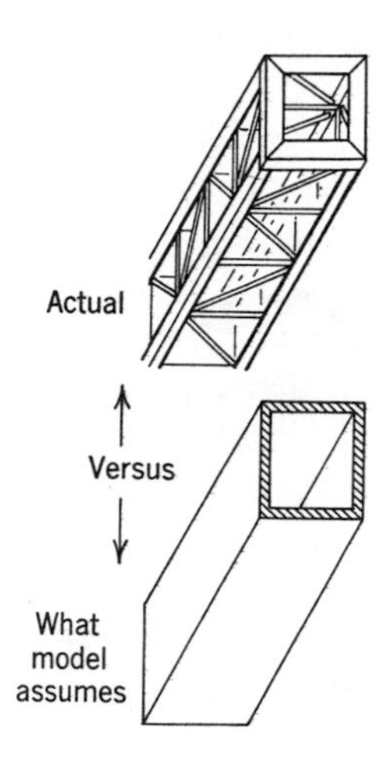

There are such discrepancies in the case of the model used to predict deflection of the Empire State antenna stack; for instance:

- **According to the model, the beam is constructed out of homogeneous material, yet it is welded out of steel angles.**
- **The wind load is assumed to be constant, or at least gradually applied, yet the winds especially at that level are very gusty.**
- **The model is based on a beam of uniform cross section over its length, yet the antenna stack tapers down in stages.**

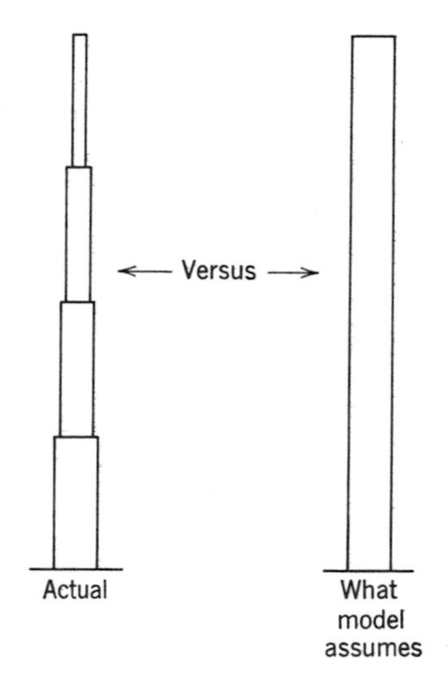

For a third example, recall the equation of state of an ideal gas; there *are* differences between an ideal gas and the real thing. *Yet these three models can provide useful predictions.* The fact that some assumptions made by a model are not consistent with reality, by itself, is unimportant. What matters in predictive modeling is whether or not the final result—the predictions—are satisfactory for the particular purpose at hand. (Who knocks the crystal ball that gives him accurate predictions?)

Simplifying assumptions are made for good reason. If some of the complicating factors are not excluded from the model, it is virtually impossible to develop a usable predictor. Furthermore, many discrepancies between model and reality are of negligible *practical* consequence; their removal makes a model more complicated and costly and yet does little to improve accuracy. To be sure, some of the discrepancies between the parking lot model

Figure 10-2 *Obviously there is something in the model of the behavior of this structure that didn't jive with reality. In this instance the only penalties were repair cost and a frightening experience for the flight test crew, so we can laugh. But it doesn't always turn out this way.*

and the real thing could be removed, but the model would take much longer to set up and run; the small increase in accuracy doesn't justify it. In every instance, complicating, costly factors that are irrelevant or inconsequential are ignored and understandably so.

Developing Predictive Models

The following procedure, summarized in Figure 10-3, is a significant one in science and engineering. To develop a satisfactory predictive model:

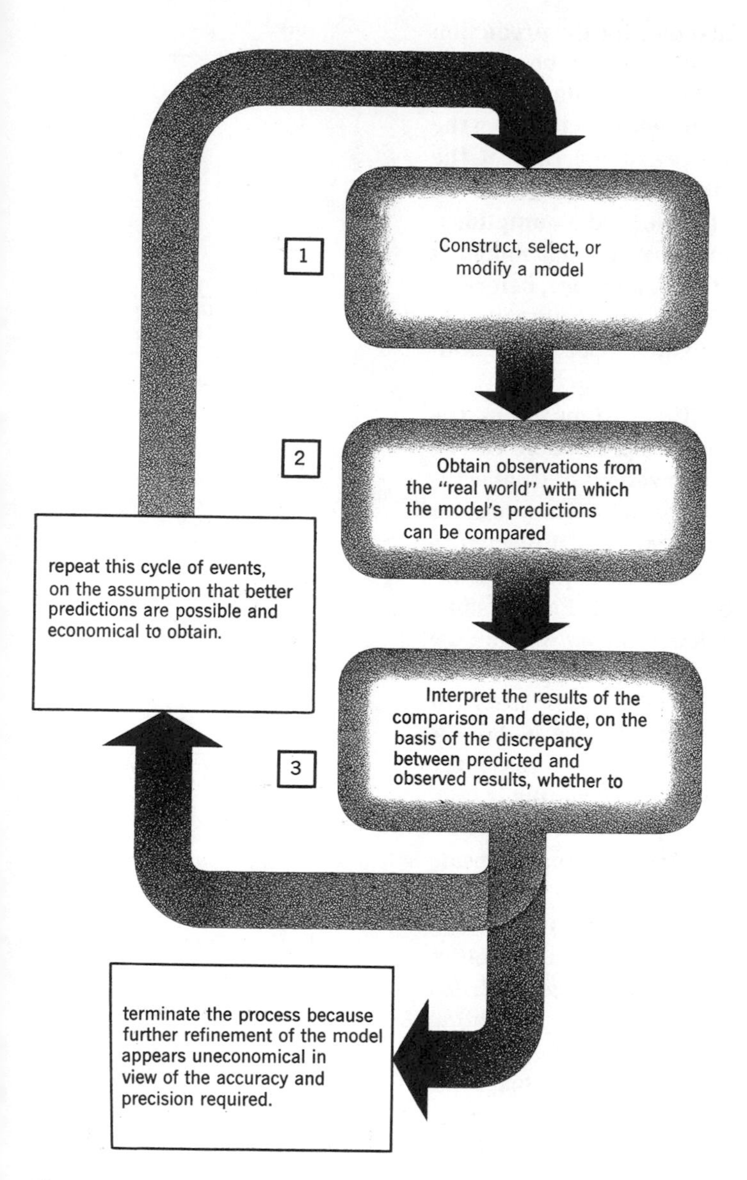

Figure 10-3 *A diagrammatic model of the general procedure for developing a predictive model.*

1. A model that is *potentially* satisfactory for the prediction task at hand is developed or selected. It can be a model prepared especially (e.g., the model for simulation of parking lot operations), or it can be one selected from the store of "ready-made" models, as in the case of the deflection equation. In the latter case, the engineer on that job is apprehensive about the violated assumptions, to the point where he questions whether that model is accurate enough in this application. Therefore, before he uses the deflection model in this and future antenna design problems, he intends to evaluate its predictive ability.
2. Therefore, he obtains some observations from the "real-world" with which this model's predictions can be compared. In this instance the observations of what actually occurs are obtained from realistic laboratory tests. A 10-foot scale model of the antenna stack is subjected to simulated wind loads of varying intensity, gustiness included, so that deflection can be measured. For the same experimental conditions, deflection is *calculated* from the deflection equation. Thus, for each set of conditions tested, he has a *predicted* deflection and *observed* deflection, which he plotted as shown in Figure 10-4*a*. The poorer a model's predictive ability is, the weaker the correlation between predicted and actual results is, and the more scattered the points in a plot like this are.
3. Now, the engineer must interpret these results and decide whether to use the model as is, refine it and repeat the process diagrammed in Figure 10-3, start over with a different model, or give up on models and turn to another means of making the predictions he needs. *In general, this decision depends heavily on the situation in which the predictions are to be used.* When the costs of errors are high, especially when life and limb are at stake, the predictions must be very accurate. In other situations fairly rough predictions are adequate. Hence the adequacy of a model cannot be evaluated independently of the particular use to which it is to be put. Therefore you should not attempt to judge the seriousness of the dispersed points of Figure 10-4*a* unless you know specifically how and where the model's predictions are to be used.

(a)

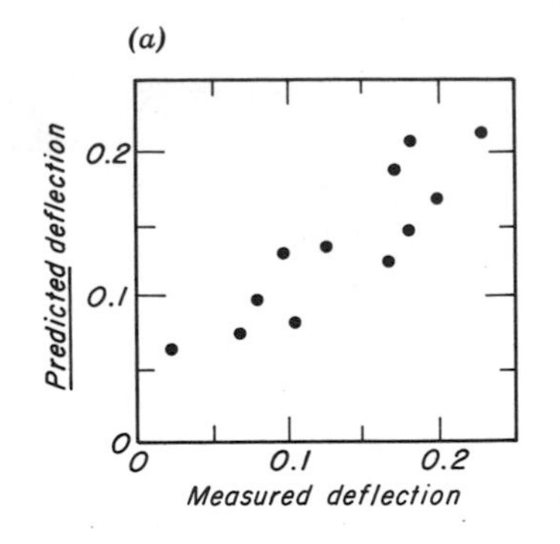

(b)

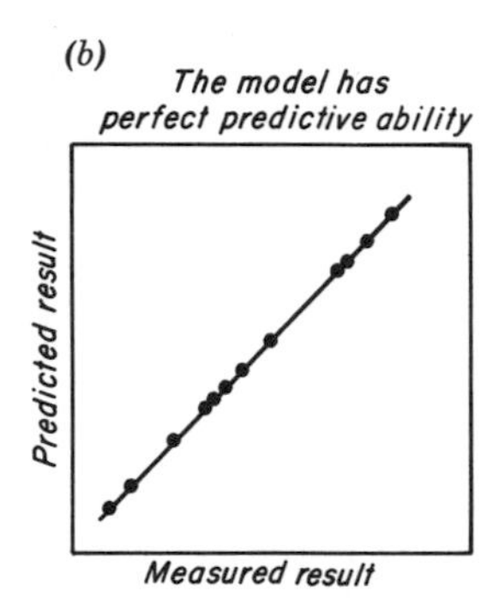

(c)

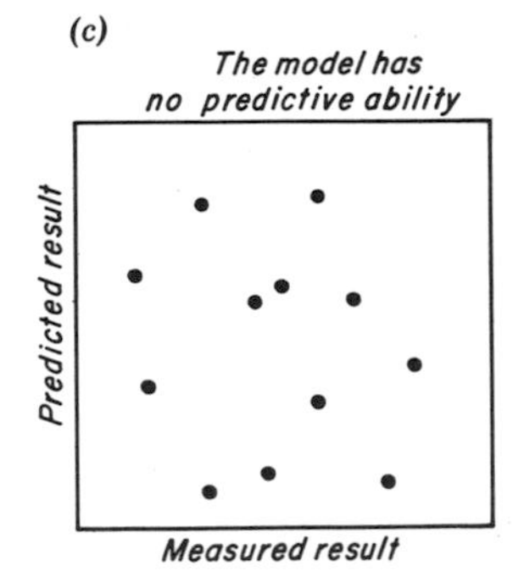

Figure 10-4

You cannot expect the perfect predictive ability demonstrated in Figure 10-4*b* for several reasons. The points will never fall on a straight line because assumptions will inevitably be violated to *some* degree whenever a model is applied. Errorless predictions are unattainable. Furthermore, there is some error in the measurement of the values substituted into the model, for example, F and E. Note, too, that predicted deflection is compared with *measured* (i.e., observed) and not actual deflection of *a scale model*. So you cannot blame all of the dispersion of points in Figure 10-4*a* on the predictive equation.

As more time is spent attempting to refine a model by the process diagrammed in Figure 10-3, the development cost of the model continues to mount. The engineer is certainly interested in this cost. He will invest no more in refinement of a model than is necessary for his purposes; remember the ''flying frame'' of Figure 1-15. Furthermore, as he works to refine a model, successive refinements become more and more difficult to achieve and less and less effective in reducing error. Thus there is a point of diminishing return on the time invested in improving a model.

Science, Engineering, and Models

The Greek model of the universe was a large, hollow sphere with lights affixed to its inside surface and with the Earth at its center. The Earth was fixed, and the sphere rotated. This was one in a succession of models of the universe that have evolved over the centuries. Man's conception of the shape of the Earth has a long and interesting history also. Of course, the flat-world model persisted for centuries. It may be way off but, until the voyages of sailors lengthened to the point where navigation based on it was seriously in error, the discrepancy didn't matter. Another fascinating story is the history of man's models of his own brain.

The history of science is replete with cases like these of models that (we know now) were very unrealistic representations of reality. Even today discrepancies exist in all scientific models, of lesser magnitude and consequence to be sure. The history of science is essentially the history of man's attempts to remove discrepancies from his models of nature. In fact, science *is* the continuing process of refining man's models of the natural world.

If you are clear on the differing roles of models in science and engineering, you are clearer on the distinction between the two fields than some writers are. To the engineer, a model is a means to an end, a tool. But in science, the model *is* the end. Furthermore, on the matter of when and how long to strive for refinement of a model, there is a significant difference between science and engineering. The engineer's basis for deciding when to terminate the refinement cycle diagrammed in Figure 10-3 is basically an economic one; as an engineer I invest no more in a model than necessary to solve my problem. A scientist has other bases for deciding when to terminate this process that are less tangible, quite elusive, and often personal. In many instances he is driven by personal interest in a subject, by dedication to a cause, by an obsession, or by an urge for perfection, to go far beyond the point I would go in refining a model. This is not to say that a scientist is oblivious of matters such as the cost of continued efforts to refine a model, of diminishing returns on those efforts, and of other scientific questions he could turn to, but the weights attached to these criteria by scientists and engineers are certainly different.

Modeling—Finale

This concludes two chapters on modeling but hardly terminates your contact with models. In fact, in the very next chapter you will gain additional insight into the roles of models in and outside of engineering. If you are familiar with the types of models introduced in the previous chapter, and if you have developed a feel for the utility of models as demonstrated in this chapter, you have assimilated something that you will almost certainly find useful on numerous occasions that, perhaps, are somewhat difficult for you to visualize at the moment.

EXERCISES

for Chapters 9 and 10

1. Develop a model for predicting the amount of time you require for various lengths and types of reading assignments. Most of the predictions provided by this model should be within plus or minus 20 percent of your actual reading time. Explain why

the actual time and the predicted time for a given assignment very rarely agree. (There are many variables that affect reading time, some of which must be recognized by your model, some of which can be ignored because their influence is negligible, and some of which should be fixed at reasonable levels, e.g., illumination. Some variables defy measurement. Examples: difficulty of the subject you are reading and the proportion of that material you are expected to retain. These unquantifiable variables you will have to rate subjectively, and in terms of only two or three broad categories at that.)

2. Scheduling is still a headache for hospitals. Varying needs of patients must be matched with varying availability of physicians and facilities like bed space and operating rooms. Those who seek better scheduling systems generally find it impractical to experiment with the real thing to test alternative schemes, and so they find a model extremely useful.

You could develop such a model and, if you did, you would probably set up a series of submodels, or generators, as postulated in Figure 10-5. Try your hand at developing the operating room model, to the extent that given a patient and the *type* of operation he requires, the model will generate an operation time for him. Specify what information you need and the instructor will supply it.

Other submodels like this could be developed to generate, for example, the number of days a patient remains in the hospital and the number of patients seeking admission each day. You might like to carry this further, even if it means assuming data.

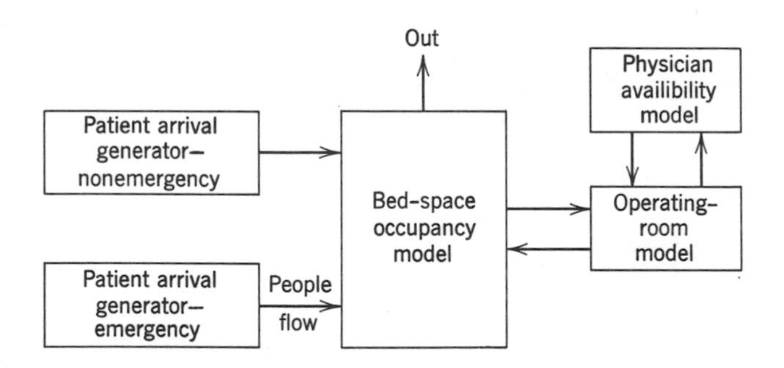

Figure 10-5 *Some of the submodels that make up a larger, more useful model of certain operations of a hospital. This illustrates how a rather large, complex model can be synthesized from a number of much smaller, relatively simple models.*

After the model is developed, it can be tied in with a scheduler, a person who makes decisions on admissions and the like, and the whole thing set into operation to test alternative scheduling policies and other proposals that might lead to improvements.

3. A large hotel caters to all kinds of banquet affairs, ranging from small receptions to large conventions. The banquet manager has a problem. If he "overprepares," food and set-up time are wasted; if he underprepares, he has problems, too. So he is interested in accurate forecasts of turnouts. Actual turnout depends on the type of affairs and numerous other variables.

(a) **Outline the steps you would take in order to develop and validate a predictive model for him. Include a list of variables that you feel *may* have to be recognized by this model.**

(b) **If so specified by your instructor, develop such a model. (Your instructor will help you with your data needs *if* you ask the right questions.)**

4. A recurring theme in the text is the relevance of a number of engineering techniques to problems that you do not ordinarily associate with engineering. Here is an opportunity to demonstrate this to yourself.

Set up and run a digital model with which you can simulate operation of the system diagrammed in Figure 10-6. You specify the data you need and your instructor will supply it.

As a minimum, develop Monte Carlo mechanisms (employing a different method in each instance) that generate daily sales and introduce real-life variation in factory replenishment time (which

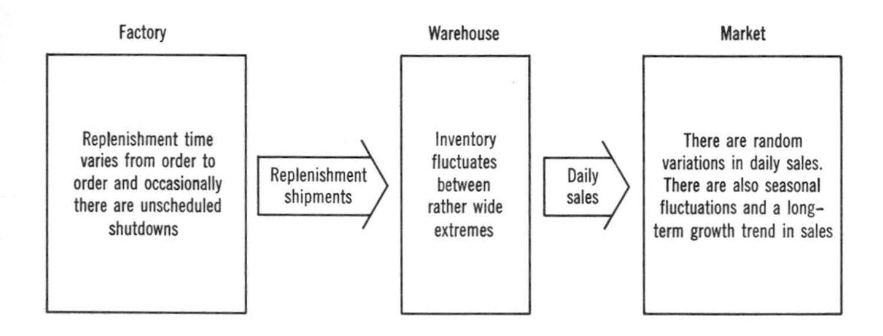

Figure 10-6 *Representation of a warehousing system showing some of the real-life behavioral characteristics of each component. The resulting inventory fluctuations characteristic of all inventory systems are characterized by Figure 10-7.*

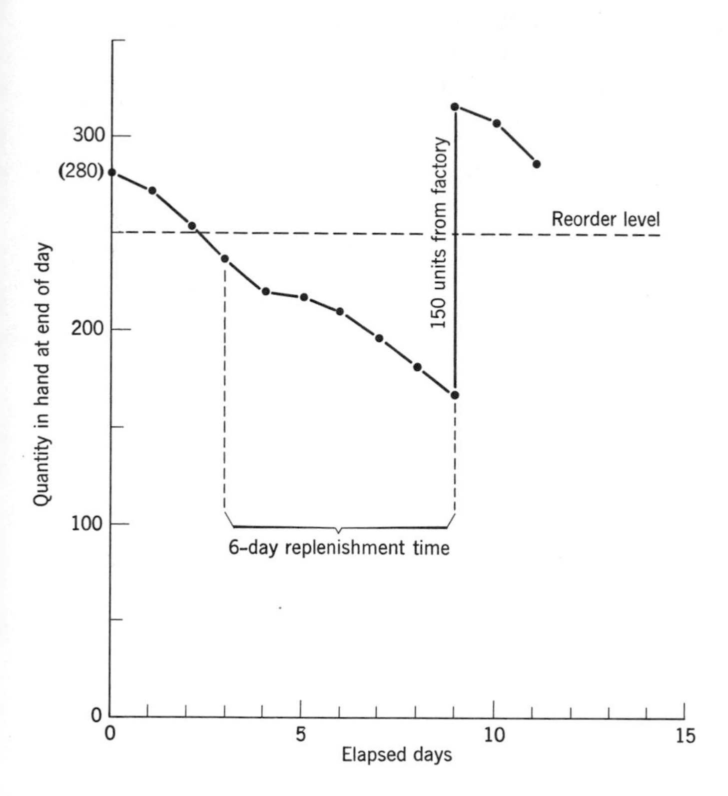

Figure 10-7

is the time lapse between the placement of an order for replenishment of warehouse inventory and arrival of the goods at the warehouse receiving dock). This enables you to synthesize a realistic inventory versus time graph, exemplified by Figure 10-7. Keep a running record of system status, using a tabulation sheet, graph, or preferably both.

Operate your model long enough to demonstrate to yourself and your instructor that you know what you are doing. *Also,* identify questions you could answer by experiments on your model if time permitted.

5. What do you envision could be learned (i.e., questions that could be answered) from a digital model of major operations at a proposed jetport? The model generates arrivals of planes and passengers, landing and take-off times, load times, breakdowns, and emergencies.

6. Obviously no predictive model perfectly represents its real-life counterpart; assumptions are never perfectly satisfied. Describe a realistic attitude toward discrepancies between assumed and true states of affairs.

7. On what assumptions is the equation of state of an ideal gas (page 141) based? Since real gases do not satisfy some of these assumptions, are the predictions provided by this model usable? Explain.

8. Choose a predictive model from another of your textbooks and identify all the assumptions that go with it. Then indicate the assumptions for which there probably is a significant deviation from the real world.

OPTIMIZATION

Chapter 11

When you adjust the focusing knob of binoculars in an attempt to sharpen the image, you are engaged in a process called optimization. Focus depends on the lens setting, which you manipulate until image sharpness is a maximum. In this and many other familiar cases there is a *criterion* which is influenced by a *manipulated variable,* and a value of the manipulated variable for which the criterion is a maximum, called the *optimum value*. The manipulated variable in each of these cases has an optimum value with respect to the criterion indicated.

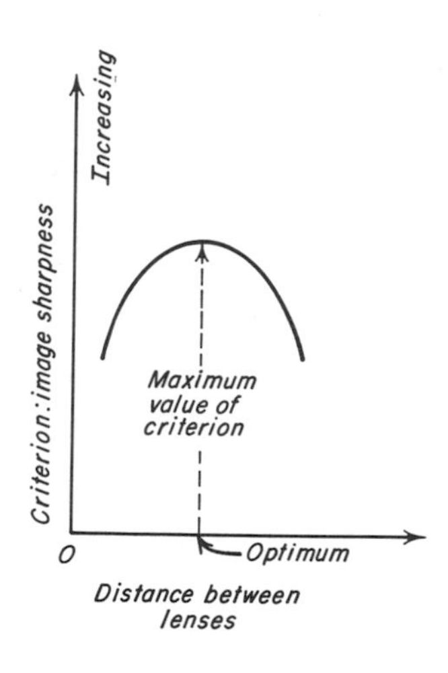

Manipulated Variable	*Criterion*
Fuel injection rate	Engine efficiency
Room temperature	Body comfort
Rate of work	Total work accomplished
Temperature	Bacteria multiplication rate

Optimum can be a minimum or it can be a maximum, so it should be viewed as *best* with respect to a given criterion, where best may mean maximization of the criterion (e.g., maximum engine efficiency) *or* it may mean minimization of the criterion (e.g., minimum with respect to cost). The concept of optimum is a basic one in engineering; there is an optimum solution to every problem. In fact, each specific characteristic of a solution has an optimum value. For example, there is an optimum size and shape for the coffee-pot handle with respect to ease of handling; an optimum process for refining petroleum with respect to total cost; and an optimum mix of ingredients with respect to the strength of concrete. Thus the concept of optimum permeates most aspects of an engineer's work. It guides his actions and decisions, serving as a goal both in the solutions he produces *and* in the way he reaches them.

Optimization is the process of seeking the optimum solution. More specifically, it is an exploratory process involving a *search for* and *evaluation of* alternative solutions in order to locate the best one. Thus, in the search and decision phases of design, the engineer is engaged in optimization.

Unfortunately, in most engineering problems, optimization is more complex and time-consuming than in the binocular-focusing case. This is so especially because of numerous conflicting criteria, which are certainly not new to you. You have dealt with them if you have tried to tune in a television program when the clearest picture and clearest sound do not occur at the same setting of the dial. The optimum settings for picture clarity (O_P) and sound clarity (O_S) do not coincide; the setting that is optimum with respect to picture clarity is *suboptimum* for sound, and vice versa. In this situation you must compromise between two conflicting criteria—picture clarity and sound clarity—conflicting in the sense that as you try to improve the situation with respect to one criterion, you are likely to make the situation worse with respect to the other. Before you can reach a satisfactory compromise, you must decide on the relative importance of picture clarity and sound clarity. Different persons will attach different degrees of importance to these two criteria, and so their compromise settings will differ, with most persons' settings falling somewhere between O_P and O_S.

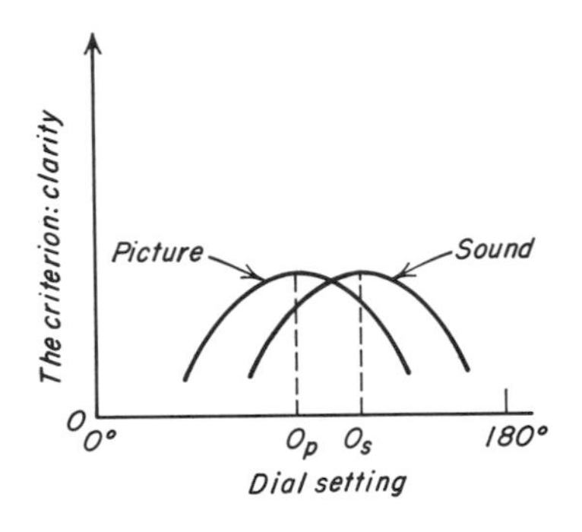

This situation—conflicting criteria and the need to find a compromise between them—abounds in engineering problems. Usually, however, the conflict is among many criteria. Take the engineer who is developing a machine for harvesting fruit. Some of the criteria he must consider are the speed with which the machine picks fruit, safety to persons near it, the degree to which it damages fruit, and cost. He cannot specify a machine that is the ultimate in speed—unless cost is no object and no one cares about safety and fruit damage, but the prospective purchasers of such machines *do* care. Hence he must sacrifice some speed to gain a reduction in cost, fruit damage, and hazards. Furthermore, he could design a machine that does not damage fruit, but it would probably be so slow and so expensive that no one would buy it. Similarly, the machine can be made perfectly safe, but only at a cost few potential buyers would consider paying; and so on for all criteria. Consequently, the engineer will alter his solution until he achieves what he believes is the optimum balance between these conflicting criteria.

This *is* a compromising process, and it gets complicated when there are more than two conflicting criteria or when they cannot be put in numerical terms. Yet it is a necessary process if the final solution is to be even close to optimum. In a relatively simple case it proceeds something like this. To determine the best balance between harvesting speed and damage to fruit, the engineer must know the relationship between these two criteria. On the basis of previous experience and some direct experimentation, this engineer estimates that damage depends on speed, approximately as shown here. With the help of this *model* the engineer can predict the reduction in fruit damage that can be achieved by a given sacrifice in speed, or what a given increase in speed will cost in terms of damage. In the course of such deliberations the engineer is determining how much speed to trade for reduced damage in order to achieve the optimum compromise. This "giving and taking" between conflicting criteria in order to reach the best balance is appropriately referred to as the *trade-off process*.

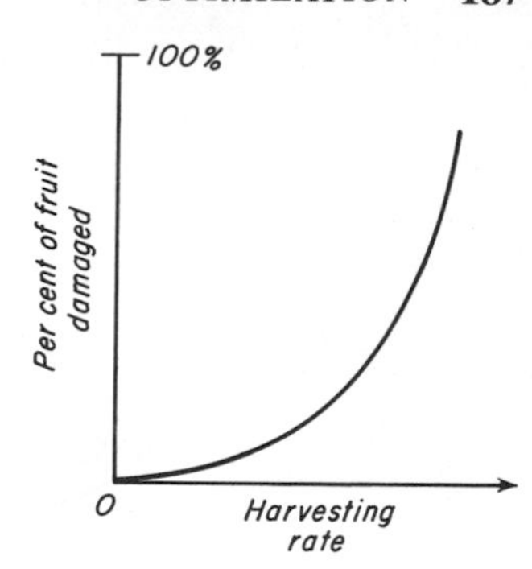

Trade-Offs

The trade-off process is all too familiar to the experienced engineer; almost all decisions involve trade-offs. The essence of this process is compromise, which is both essential and difficult. Some examples follow.

In the old days an aircraft penetrating enemy territory was inclined to fly as high as possible to (hopefully) remain out of range. But today, with radar and antiaircraft missiles, the smartest tactic is to remain as low as possible. This has given rise to a new art for military pilots: terrain following—flying as low as possible, hugging the contour of the land. It also gives rise to the rather delicate trade-off situation graphed in Figure 11-2.

Recall the trestle of the bridge-tunnel described in Chapter 1. The choice of span between supporting piers illustrates the trade-off process and is also an interesting example of optimization in design. Figure 11-3 tells the story.

You can visualize some of the trade-offs that must be made in design of a mass transportation system. Criteria like convenience, speed, load-carrying capacity, and construction cost are conflicting and, in some instances, unquantifiable.

Value Decisions

The engineer cannot arrive at the optimum trade-offs between criteria until he has estimates of the relative importance of each

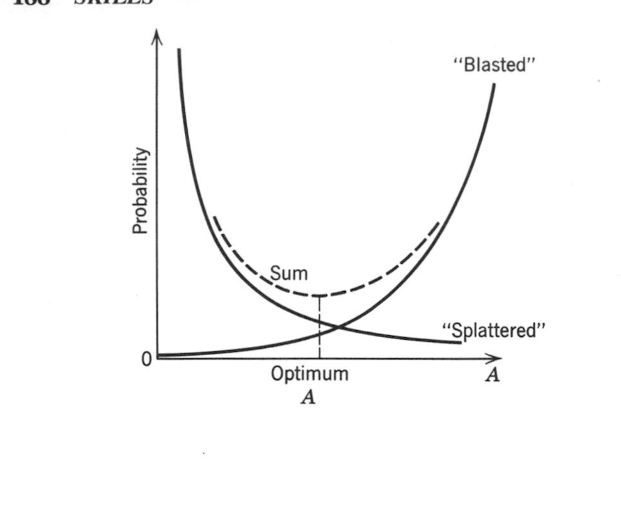

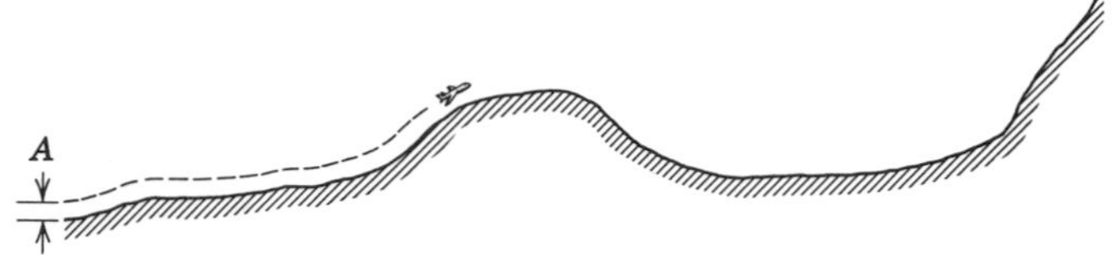

Figure 11-2 *As the pilot flies closer to the ground (i.e., as A decreases), the probability of splattering himself on the side of a mountain increases. On the other hand, the higher he flies, the greater the probability of being detected and blasted out of the air with a missile. So he has a trade-off to make, which minimizes the sum of these two probabilities and, therefore, the probability of "getting it" one way or the other.*

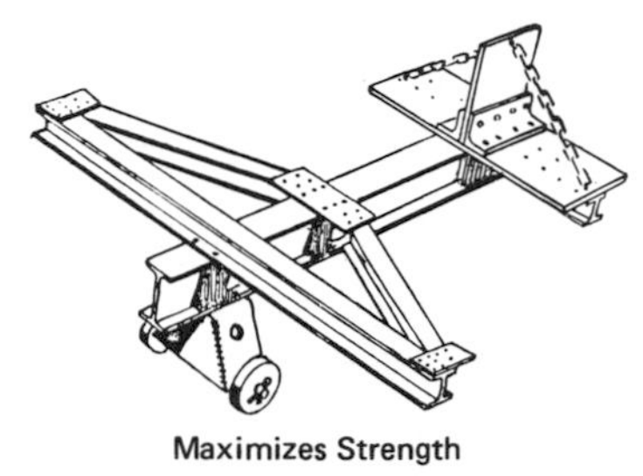

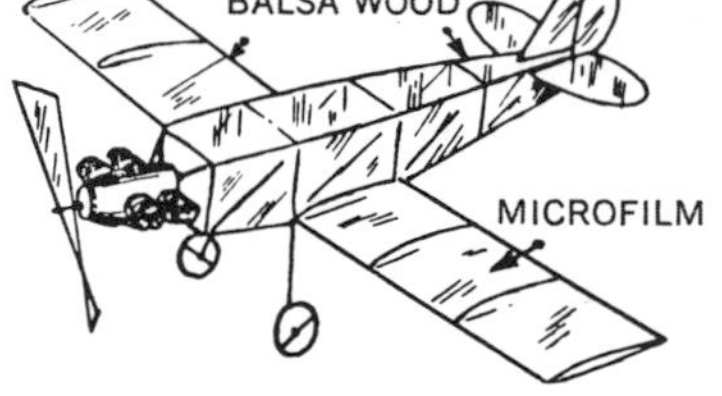

Figure 11-1 *One reason that you don't see planes like these is that trade-offs are made in aircraft design in order to achieve an optimum compromise among conflicting criteria.*

(recall the television tuning example). Knowing the relationship between harvesting speed and damage to fruit is of limited help until the engineer learns the relative importance of different speeds and degrees of damage to those who are potential users of this machine. Perhaps most users attach considerable value to high harvesting speed and are willing to tolerate a rather high percentage of damaged fruit in order to get this speed. Once the engineer knows the relative values attached to criteria, he is in a position to make trade-offs.

The assigning of relative value (importance, weight) to a criterion is commonly called a *value decision* (some say value judgment). Value decisions in engineering are difficult to make, partly because the engineer must anticipate the value attached to a given criterion *by others,* often a large group at that. Designers of a new-model automobile are contemplating the addition of a new

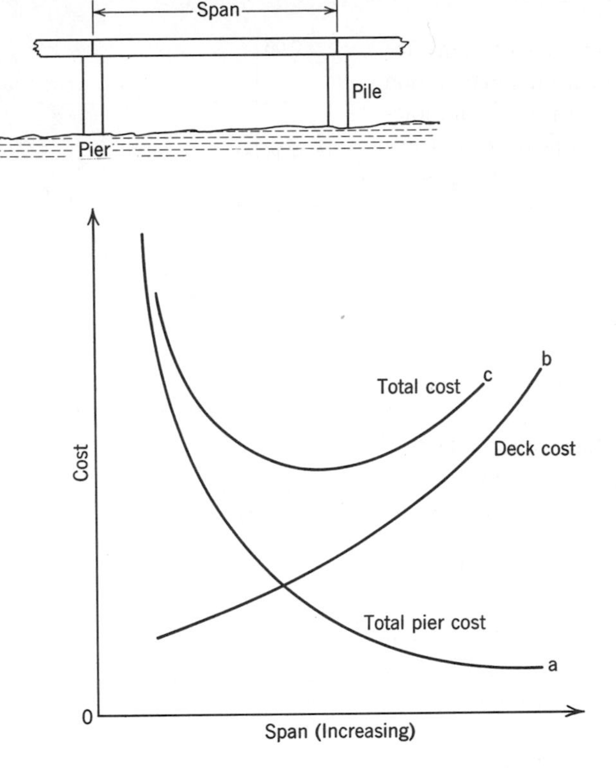

Figure 11-3 *As indicated by curve a, if the distance between piers is increased, the total cost of piers diminishes, since fewer of them are needed for a given crossing. But there is a price to be paid as the span is lengthened: the roadway deck slabs must be sturdier, so their cost goes up, as indicated by curve b. It's another trade-off situation. The optimum deck length is determined by the minimum point on the total cost curve c.*

safety feature. They must anticipate the relative values that potential buyers will attach to reduced risk of serious injury as opposed to *X* dollars this device will add to the price of the automobile.

Value decisions are never more difficult than when they involve human life. The designer of a highway must consider criteria such as construction cost and safety which, of course, are in conflict. He could specify a virtually impenetrable medial barrier for a four-lane dual highway under design, thereby adding $750,000 to total construction cost and reducing fatal accidents by two thirds. Is the added safety worth the cost? Ideally, taxpayers should answer that question for the engineer, but it is hardly feasible for him to interview a large sample of taxpayers every time a decision like this is to be made. The only practical alternative for him is to second-guess the public's preference.

This "dollars versus life" dilemma is only one of many instances in which criteria cannot feasibly be measured in common units. Dollars are convenient units, but many criteria are impractical to put in monetary terms—safety being a conspicuous example. Yet safety is a criterion to be reckoned with in most engineering problems. The inevitable result is that an engineer has some very trying value decisions to make.

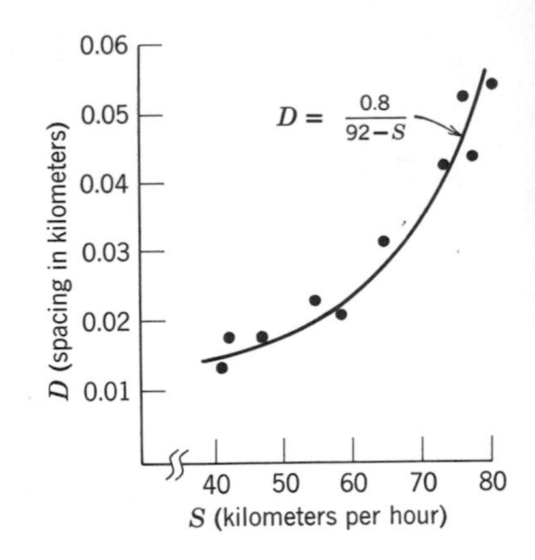

Figure 11-4 *The engineer on this project obtained measurements of vehicle speed and spacing between successive vehicles from electronic apparatus set up in one of the city's tunnels. Each point on this graph represents a number of observations and can be interpreted as follows: for vehicles traveling around 70 kilometers/hour, the average spacing is 0.036 kilometers.*

Optimization Procedure

A traffic engineer is making a study with the objective of maximizing the number of vehicles that can pass through New York City's heavily traveled tunnels. (No, this is not simply a matter of speeding up the vehicles.) You know and so does he that drivers increase the spacing between vehicles as their speed increases. In fact, he has data (Figure 11-4) for a large sample of drivers indicating that spacing increases at a faster rate than speed, which leads him to suspect that encouraging drivers to move faster will increase throughput (vehicles per hour), but only up to a certain speed. He expects that beyond that point, encouraging drivers to move faster will actually reduce the throughput because increased spacing cancels out the effect of increased speed. If his suspicion is correct, it follows that there is some average vehicle speed that is optimum (i.e., results in maximum throughput). He tested this hypothesis, proceeding in the following manner.

He developed the following mathematical model from his spacing versus speed data (Figure 11-4) by finding the equation describing the relationship between D (vehicle spacing) and S (speed) indicated by that data.

$$D = \frac{0.8}{92 - S} \qquad \text{(a)}$$

In order to convert this model into a form that describes the effect of S (the manipulated variable) on the criterion C (throughput in this case), he used

$$C = \text{throughput} = \text{vehicles per hour}$$

$$= \frac{S}{D} = \frac{\text{kilometers per hour}}{\text{kilometers per vehicle}} \qquad \text{(b)}$$

and substituted equation (a) into equation (b) thusly:

$$C = \frac{S}{D} = \frac{S}{\frac{0.8}{92 - S}} = 115S - 1.24S^2 \qquad \text{(c)}$$

This mathematical model is called a *criterion function* (sometimes *objective function)*. It describes in what way the criterion is a function of one or more manipulated variables.

To determine what value of S maximizes C (i.e., what vehicle speed will get the most vehicles through the tunnel per interval of time), he substituted a trial series of speed values into equation (c) and computed the resulting values of C. The results are shown in Table 11-1 and Figure 11-5. S values in multiples of five were used to determine the general location of the maximum C, which

TABLE 11-1 GLOBAL SEARCH

Trial Values of S in Kilometers per Hour	Resulting Values of C in Vehicles per Hour
35	2494
40	2600
45	2644
50	2625
55	2544

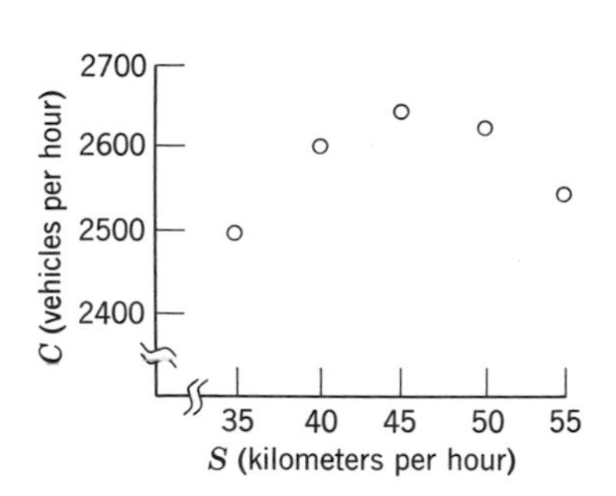

Figure 11-5 *Global search, based on Table 11-1.*

TABLE 11-2 LOCAL SEARCH

Trial Values of S in Kilometers per Hour	Resulting Values of C in Vehicles per Hour
41	2614
42	2625
43	2634
44	2640
45	2644
46	2645
47	2644
48	2640
49	2634

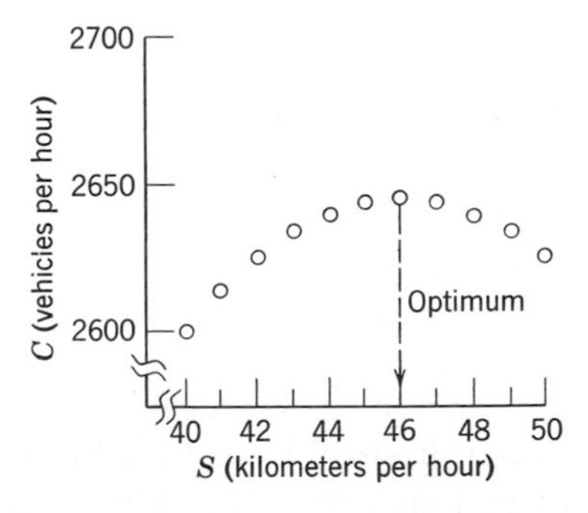

Figure 11-6 *Local search, based on Table 11-2.*

appears to be near 45 kilometers per hour. Then he explored a localized range of interest by substituting S values in increments of 1 kilometer per hour, enabling him to pinpoint the optimum S at 46 kilometers per hour (Table 11-2, Figure 11-6).

The same conclusion can be reached by applying elementary calculus. Equation (c) is differentiated with respect to S, with this result:

$$\frac{dC}{dS} = 115 - 2(1.25)\,S = 115 - 2.5\,S \qquad \text{(d)}$$

dC/dS is the rate at which C changes as S changes and, therefore, it is the slope of the curve in Figure 11-7. It is apparent that this slope is different at different values of S. At the peak of the curve, where C is a maximum, this slope is zero. Thus it remains to solve for the value of S at which

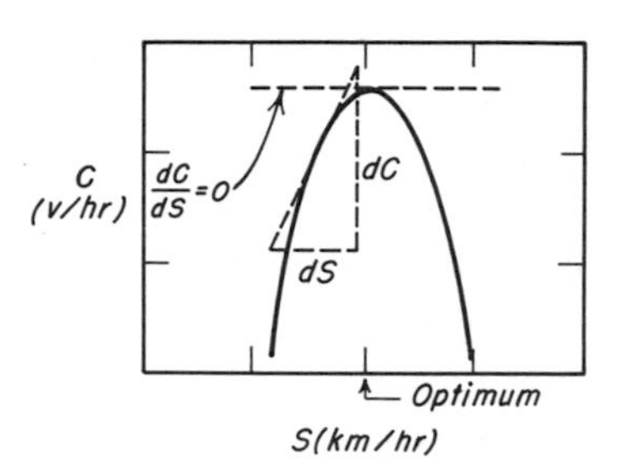

Figure 11-7

$$\frac{dC}{dS} = 0 \qquad \text{(e)}$$

But

$$\frac{dC}{dS} = 115 - 2.5S$$

So

$$\begin{aligned} 115 - 2.5S &= 0 \qquad \text{(f)} \\ S &= 46 \end{aligned}$$

Therefore, the optimum S is 46 kilometers per hour.

Iterative Optimization

The preceding two means of arriving at the optimum vehicle speed illustrate basically different methods of locating an optimum solution. One of these is the *iterative method* and it generally goes like this, using Figures 11-5 and 11-6 to illustrate.

I. The *global* search.
   A. For the manipulated variable the engineer assumes several values spaced over a range he suspects includes the optimum. (Five vehicle speeds over the 35 to 55 kilometers per-hour range were selected.)
   B. He predicts the effect of each of these assumed values on the criterion. [Equation (c) was used for this purpose, generating Figure 11-5.]
II. The *local* search.
   A. Benefiting from the global search, more values of the manipulated variable are selected but over a narrower range. (Since Figure 11-5 indicated that the optimum is in the vicinity of 45 kilometers per hour, he chose values around it for investigation.)
   B. The effects on C are again predicted. [Using equation (c) he generated Figure 11-6.]

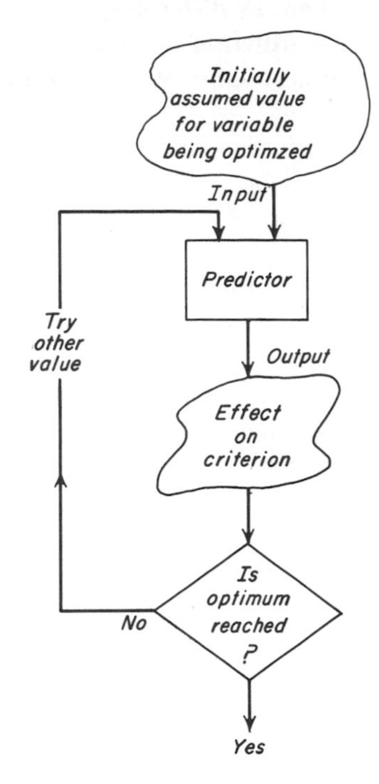

Figure 11-8

Ordinarily this two-phase search will permit the engineer to estimate the optimum value for a manipulated variable satisfactorily. If not, the above procedure is repeated over a still narrower range.

The iterative method of optimization is basically an accelerated learning process. Through a series of successive approximations, the engineer gradually "closes in" on the optimum value of a manipulated variable, as diagrammed by Figure 11-8.

Analytical Optimization

The other of the two basic procedures for locating an optimum is the *analytical method*. In this case a mathematical model yields the optimum value directly, as illustrated by the second method of arriving at the optimum vehicle speed in tunnels. Often the model is derived through the use of calculus, generally proceeding as follows.

1. **A criterion function [e.g., equation (c)] is developed. Its general form is**

$$C = f(V)$$

where C is the criterion and V is the manipulated variable. Although C is often measured in dollars, it need not be. It could be number of passengers carried, pounds, or efficiency.

2. Then, by differential calculus or other means, the criterion function is converted to a form that yields the optimum value of the manipulated variable directly.

The criterion function can contain more than one variable to be optimized, in either the iterative or the analytical method. The general form for the criterion function, then, is

$$C = f(V_1, V_2, V_3, \ldots)$$

where V is a manipulated variable.

Surely you wonder why anyone would use the iterative procedure, considering the directness of the analytical method. It's a good question and there is a good reason: in many cases the analytical method is much too difficult mathematically and the iterative method is the only practical way. Furthermore, in many instances the engineer wishes to know how *sensitive* the criterion is to deviations of a variable from its optimum value, and the iterative method ordinarily provides this information conveniently. Note from Table 11-1 that very little is lost in terms of vehicles per hour if the average speed should deviate slightly from the 46 kilometers per hour optimum. If the average speed is 50 kilometers per hour, there is a loss of only 20 vehicles per hour, which is a drop of less than 1 percent. This information is useful because it would be difficult to attempt to control the average speed to 46 kilometers per hour and so the engineers were interested in learning what sacrifice in vehicles per hour would be made if they attempted to control speed to a higher value. An investigation of this kind, to learn the consequences of setting a variable at something other than its optimum value, is a *sensitivity analysis*.

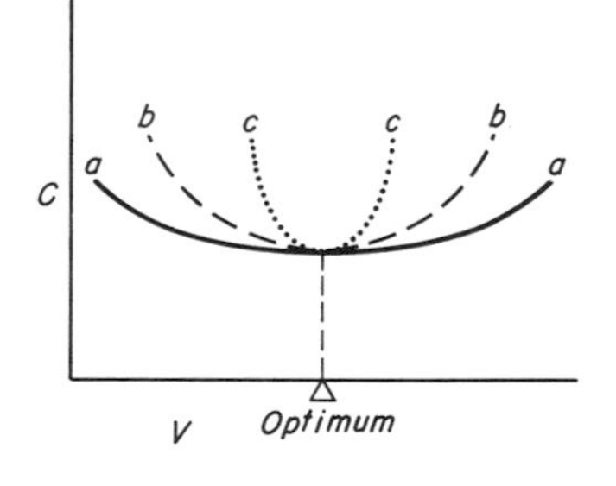

The formal optimization methods illustrated above cannot be relied on as much as the engineer would like. Because of a preponderance of unquantifiable criteria, the large number of variables, or a lack of time, engineers often must rely on procedures that are less formal, less quantitative, and less objective than those illustrated. Usually they use a variety of different methods in their optimizing efforts, the particular combination of procedures varying considerably from problem to problem (so much so that it is foolish for an author to attempt to generalize any more than I have).

Optimization of Problem-Solving Methods

The concept of optimum, applied up to now only to solutions to problems, is just as applicable to the *methods* the engineer employs in arriving at those solutions (e.g., the measurement systems, the computational methods, the models, and the number and types of technicians utilized).

Over the long run the errors in a model's predictions cost something. This is so because of mistakes, failures, accidents, repairs, and alterations that result when decisions are based on erratic predictions *or* because of the high safety factors necessary to protect against such adverse occurrences. And yet it is worth reducing these errors through model refinement only up to a point. For instance, the designers of a chemical plant are relying on a model to make predictions on which to base their design. If, after the plant has been built, there is a small disagreement between performance predicted and performance experienced, it is of little practical consequence and is considered inevitable. In this instance the cost of the lack of correlation between predicted and actual results is negligible, as indicated by point 1 on curve *a* of Figure 11-9. A larger discrepancy, however, might well result in some wrong decisions that are discovered only after the plant is built, the penalty for which is some costly alterations. In this case the cost of the lack of correlation may be somewhere around point 2 of curve *a*. A still larger lack of correlation could result in something much more costly, an explosion, the cost of which might be in the area of point 3, (refer to Figure 10-2, page 176). Curve *a* indicates what generally happens to the cost of errors in a model's predictions as the engineer refines the model and reduces those errors. The cost declines, but at a decreasing rate, and so a point is eventually reached at which additional refinements are of negligible benefit and not worth striving for.

There is another good reason for not attempting to refine a model to the point where curve *a* levels off: the cost of developing and applying it increases, as indicated by curve *b* of Figure 11-9. As additional efforts are expended to reduce the error in a model's predictions, this cost accelerates upward because additional improvements become more and more difficult and time-consuming to achieve. This is generally true.

The optimum degree of refinement, then, is the point at which the sum of these two costs is a minimum, as shown by curve *c*. It is uneconomical to attempt to refine this model beyond that point; likewise for all models. Therefore, there is an optimum degree to which a model should be refined.

The situation summarized by curve *c* is not unique to modeling. The same holds true for measurement systems, information-search methods, and most tools and techniques. In each case there is an optimum degree of refinement simply because the man-

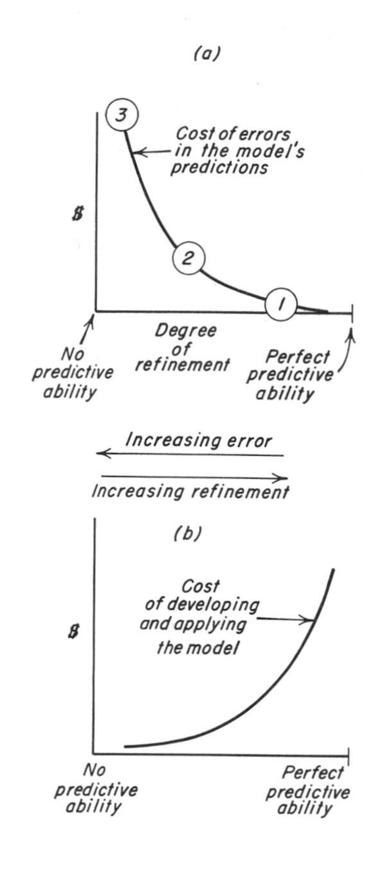

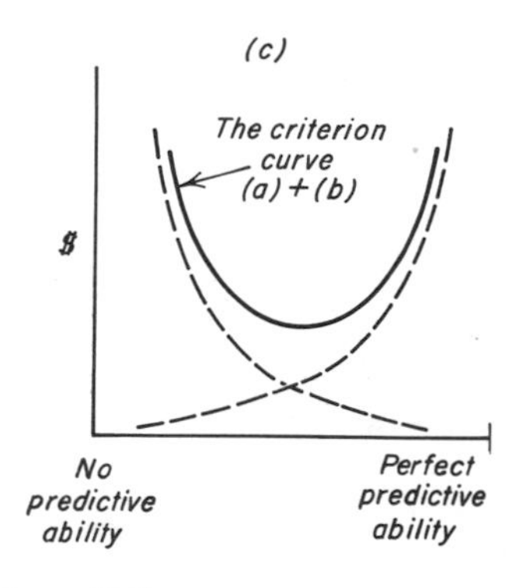

Figure 11-9

power and other resources devoted to such efforts are valuable for other purposes and because continued refinement efforts bring diminishing returns. For these reasons there is an optimum number of man-hours to devote to the solving of a problem.

Optimum As a Goal

An important goal in engineering endeavors is that of the "optimum." Engineers strive to produce optimum solutions to problems and to do so by optimum means. Notice the word *strive*. Although the optimum solution is almost always an objective, it is seldom a realization. Many real-world problems are too complex for an optimum solution to be located in a reasonable period of time. In many instances, the amount of time required would be longer than the life of the problem. Invariably, many other problems are awaiting attention, so that it often becomes more profitable to turn to one of these than to continue searching until optimum is achieved for the current problem. Thus, in design, it is usually a matter of progressing toward the optimum, continually seeking better solutions until the effort becomes more profitably spent elsewhere.

APPLICABILITY OF OPTIMIZATION CONCEPTS AND TECHNIQUES

You can capitalize on this brief introduction to optimization simply by calling on your knowledge of algebra, graphs, and other relatively elementary aspects of modeling; you can do even more with some knowledge of calculus. These concepts have a surprisingly broad range of applications, which includes a variety of nontechnical problems. Here is an illustration of how you can apply the contents of this chapter to a problem you do not associate with engineering.

Periodically the Neversmudge Printing Company must replenish its supply of paper. The frequency with which it does so and the quantity of paper it buys at one time significantly affect the cost of their paper. For instance, each time they replenish their paper stock, certain costs are incurred due to paperwork, handling, and other activities. The total of these costs, the *acquisition cost,* is relatively constant, regardless of the amount of paper ordered. This is $25 per purchase in Neversmudge's case.

Of course, it costs money to store the paper, since it has to be housed and insured, and because of interest on the money tied up in inventories. The total of these costs is called the *storage cost*. In Neversmudge's case this amounts to \$4 per year for a roll of paper.

If they order one roll at a time—as needed—they can keep the total storage cost to a bare minimum but, in doing so, they will inflate the total acquisition cost ridiculously. Or they can order a whole year's supply of paper at once, which minimizes the acquisition cost but results in a prohibitively high yearly storage cost. Somewhere between these extremes is a "compromise purchase quantity," which results in minimum *total* cost to the company. Knowing something about optimization, calculus, and graphs, *you* could find the optimum purchase quantity for them, by the following procedure.

First, you know that the total annual cost in this case is

$$T = \text{(yearly acquisition cost)} + \text{(yearly storage cost)*} \qquad \text{(g)}$$

$$= \begin{pmatrix}\text{number of orders}\\ \text{placed in a year}\end{pmatrix}\begin{pmatrix}\text{acquisition}\\ \text{cost}\end{pmatrix} + \begin{pmatrix}\text{number of rolls}\\ \text{ordered per}\\ \text{purchase}\end{pmatrix}\begin{pmatrix}\text{storage}\\ \text{cost}\end{pmatrix}$$

$$= \left(\frac{\begin{matrix}\text{rolls of paper}\\ \text{needed per}\\ \text{year}\end{matrix}}{\begin{matrix}\text{number of rolls}\\ \text{ordered per}\\ \text{purchase}\end{matrix}}\right)\begin{pmatrix}\text{acquisition}\\ \text{cost}\end{pmatrix} + \begin{pmatrix}\text{number of rolls}\\ \text{ordered per}\\ \text{purchase}\end{pmatrix}\begin{pmatrix}\text{storage}\\ \text{cost}\end{pmatrix}$$

Since algebraic manipulations in terms of words are awkward to say the least, surely you would convert the above expressions to symbols like these before proceeding.

N = number of rolls of paper required per year (3600 in this instance)
A = acquisition cost per order (\$25 in this case)
S = storage cost, in dollars per roll per year (\$4 in this instance)
Q = the number of rolls per purchase, called the purchase quantity (which you are seeking to optimize)

* In this example the cost of the material itself is ignored, since it is the same for all purchase quantities and therefore irrelevent to this decision.

This enables you to express equation *(g)*, the criterion function, as

$$T = \left(\frac{N}{Q}\right) A + (S \times Q) \qquad \text{(h)}$$

You also know that equation (h) can be differentiated with respect to Q,

$$\frac{dT}{dQ} = \frac{-N\,(A)}{Q^2} + S \qquad \text{(i)}$$

and that you can find the value of Q for which T is a minimum by the same method outlined earlier.* And so:

$$\frac{dT}{dQ} = \frac{-N\,(A)}{Q^2} + S = 0 \qquad \text{(j)}$$

$$Q = \sqrt{\frac{N \times A}{S}} \qquad \text{(k)}$$

Thus, you have found a simple algebraic expression, equation *(k)*, that yields the *optimum* purchase quantity (i.e., the purchase quantity that minimizes total cost). By substituting numerical values, you calculate the optimum Q value as

$$\sqrt{\frac{3600\ (25)}{4}} \quad \text{or 150 rolls}$$

Bravo! In a few minutes of pencil-and-paper work you have the answer for the manager of purchases. He is impressed, but he is also skeptical, since he doesn't understand calculus. But you know how to get around that; you prepare Table 11-3 and Figure 11-10 for him.

This case illustrates the types of problems that *you* are equip-

* When $dT/dQ = 0$, T is either at a maximum or a minimum. In this instance you are familiar with the function and happen to know that equation (j) represents a minimum value for T. However, in the absence of such knowledge, you could test for maximum or minimum by taking the second derivative of equation (i) and determining whether it is positive or negative at the point in question.

TABLE 11-3 DETERMINATION OF OPTIMUM PURCHASE QUANTITY BY THE ITERATIVE METHOD

Q	$\frac{N}{Q}$	$\left(\frac{N}{Q}\right)A$	$S(Q)$	$T = \left(\frac{N}{Q}\right)A + S(Q)$
Number of Rolls per	Number of Purchases per Year	Yearly Acquisition Cost	Storage Cost per Year	Total Yearly Cost
600	$\frac{3600}{600} = 6$	6 × \$25 = \$150	600 × \$4 = \$2400	\$150 + \$2400 = \$2550
200	18	450	800	1250
150	24	600	600	1200
100	36	900	400	1300
50	72	1800	200	2000
(a)	(b)	(c)	(d)	(e)

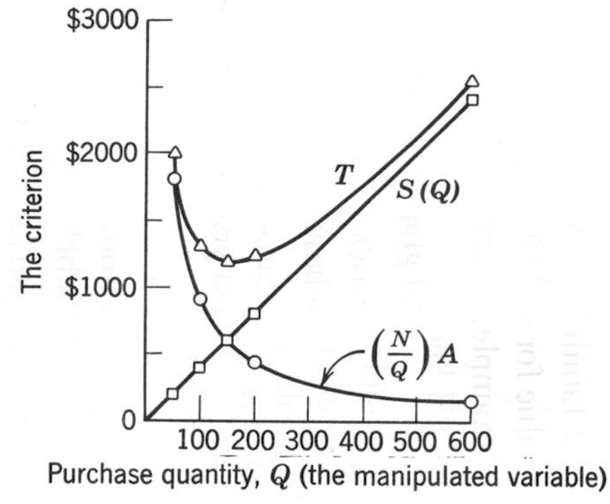

Figure 11-10 *A graphic view of columns* c, d, *and* e *of Table 11-3.*

ped to handle, with or without a knowledge of calculus. You can prove this to yourself by completing the end-of-chapter exercises.

Optimum, trade-off, and optimization should come to mind when you think and talk about some of the major techno-social problems of the times. Certainly there are major trade-offs to be made in solving environment and energy problems, trade-offs that are occasionally overlooked in the demands and debates associated with these issues. Can persons who speak of "zero air pollution" be conscious of the trade-off that must be made between cost of abatement and cost of polluted air? True, they are not always made explicit, even recognized, but trade-offs pervade political, personal, business, military, medical, and all other forms of decision making. So your encounter with trade-offs and other aspects of optimization is hardly going to terminate with the completion of this chapter.

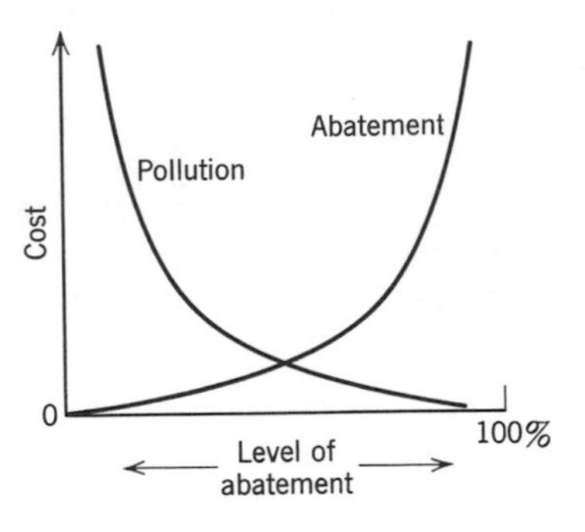

EXERCISES

1. For any two of the following, describe what you believe are important criteria that the designers must have considered. Identify criteria that appear to be conflicting and, therefore, between which trade-offs probably had to be made.

(a) *An interchange between two intersecting dual highways.*
(b) *An artificial hand.*
(c) *An automobile.*

2. Cite 10 familiar situations in which there is obviously an optimum value for some variable with respect to a stated criterion. (For example, there is an optimum reading speed with respect to total knowledge assimilated.)

3. Using a table and graph patterned after those on page 199, find the number of cases of maraschino cherries that Friendly Markets should purchase whenever it replenishes warehouse stock. Assume an acquisition cost of $50 per order, a storage cost of 50¢ per case per year, and an annual consumption of 5000 cases.

4. Using the graph from the preceding problem as the basis of your estimate, how far below the optimum purchase quantity can Friendly Markets go and raise total annual cost by only $50? How

much can they increase this quantity without raising total annual cost more than $50? This is a form of *sensitivity analysis*. Explain its usefulness.

5. In laying telephone cable along the ocean floor the chances of breakage by anchor snags and the like are reduced by allowing some slack in the cable. But cable costs money—plenty of it, in fact, so that you might think twice about "wasting" any cable. Being the perceptive creature you are, you immediately identify this as an optimization problem. Determine the optimum amount of slack by graphical means, assuming the following:

- **D = linear miles of ocean traversed (Figure 11-11).**
- **L = length of cable used.**
- **S = slack = $L - D$.**
- **That a new cable installation is under design, where D = 80 kilometers, for which cable cost is $4000 per kilometer, and for which the breakage cost (i.e., cost of interrupted service and repairs due to breaks over the life of the cable) depends on slack; this function is estimated to be**

$$\textbf{Breakage cost} = \frac{\$1{,}000{,}000}{S^2}$$

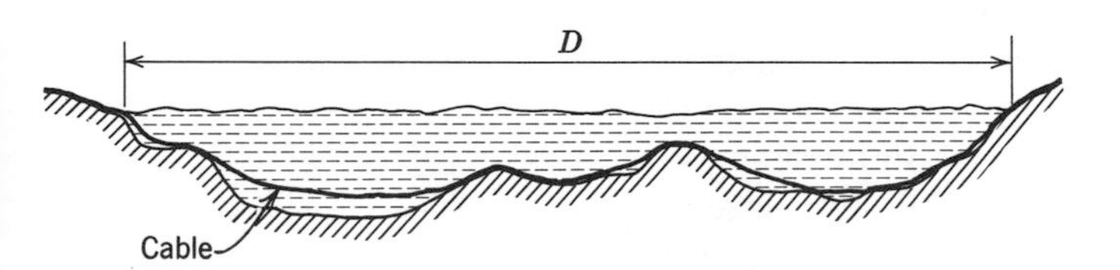

Figure 11-11

Plot *breakage cost* as a function of slack, *cable cost* as a function of slack, and then *total cost*. Your instructor will expect you to use good judgment in your choice of scales and consideration of him in your labeling and titling.

6. YOUR OPPORTUNITY TO HELP THE SAMBERIANS! Sometime during the nineteenth century, a Por-

tuguese trader fleet coming around the Horn from Asia was shipwrecked, and its entire cargo of bamboo poles was washed up on the beaches of Samberia. The native Samberians, not knowing what to do with bamboo but thinking it valuable, carefully stored all the bamboo poles in dry caves located in the interior.

During World War II, the U.S. Air Transport Command maintained a refueling stop in Samberia, leaving behind vast numbers of oil drums.

Water supply is a problem in Samberian villages and you, as a Peace Corps worker, have a brilliant idea: make stands out of bamboo to support the drums about six feet off the ground, to provide water supply tanks for village families.

The stands must support the load and still conserve the limited supply of poles. What design do you recommend?

In a period of time prescribed by your instructor, you are to design a tower (do not run to the library and research it), then to construct and test a scaled-down version. A can approximately 10 centimeters in diameter and 18 centimeters high can represent the drum, balsa strips 0.159 centimeters (1/16 inch) square can represent bamboo, and glue can be used to replace vine. Test your 20-centimeter tall model by filling (you hope) the can with water. Since the winds of Samberia could topple such a tower or vibrate it to death, your instructor will subject any model that supports a full can of water to "the fan test." And if your model passes that, he will probably get you with the "elephant rub" test. (Elephants of Samberia are known to cure an itch by rubbing against such objects!) Good luck.

Note that there are at least three criteria at work here: strength, materials consumed, and time-to-accomplish. The main trick here is to come up with a structure that supports the prescribed load, yet consumes a minimum of material and, furthermore, to do so under pressure of time. Clearly this is an optimization problem in which there is a key trade-off to be made between strength and materials consumed.

7. In what way is the tanker simulation model (problem 2, page 162) being employed in the larger process of optimization?

8. Trade-off situations are prevalent indeed. Think about the following circumstances and then record some of the trade-offs that must be made by the decision maker(s).

(a) *Energy supply versus environmental quality.*
(b) *Small college versus a large university.*
(c) *Human freedom versus the "right" to a reasonably safe and healthy environment.*

9. Referring to problem 1, page 162, find the optimum number of machines to assign to each attendant if the man's cost is $5 per hour and the machine's is $15 per hour.

10. Recall the ''banquet turnout model'' referred to in problem 3, page 182. In words, how would you employ such a model to arrive at the optimum planning policy for the banquet manager?

11. Prepare a table summarizing the loss in throughput and saving in travel time for average speeds of 50, 60, 70, and 80 kilometers per hour, based on the illustration starting on page 190. Show percentage and absolute change with respect to the 46 kilometer per hour optimum. (The tunnel is 2.3 kilometers long.) What do you conclude? On the basis of these results, do you have anything to recommend to the tunnel management? What trade-off is involved?

Knowledge Employed by Engineers

PART 3

As important as the skills of Part 2 may be to an engineer, to create complicated machines and structures, he must have something else: theory and practical know-how. The next three chapters offer a sampling of the knowledge that an engineer must have and some generalizations about it.

CONTROL SYSTEMS

Chapter 12

You have known for a long time that engineering education includes subjects like chemistry and physics and, in fact, you probably know a fair amount about these subjects. Beyond that, however, your picture of what engineers must know in order to create complex systems is probably very fuzzy. This will be remedied in part by the following introduction to the important subject of feedback control systems. Besides offering further insight into engineering, exploration of this subject provides you with some general knowledge of unusual importance. Control systems have widespread significance in social, political, biological, and economic systems, as well as in engineered systems. Thus, what follows is an introduction to control systems in general, followed by an elaboration on how this knowledge is employed by engineers.

It is impossible to aim your automobile down the highway, take your hands off the wheel, and get very far without something exciting happening. This is so even on a straight stretch mainly because there are crosswinds, slopes, and bumps that force the car off course. What is needed is a means of *detecting* drift of the car toward either side and a way of *correcting* for that drift as soon as it is detected. The means for accomplishing this are familiar: your eyes enable you to detect the need for corrective action; your brain and especially your reflexes enable you to decide what remedial action is called for; your arms and the steering mechanism are employed to implement those decisions. These parts of you and the automobile constitute what engineers call a feedback control system, the general form of which is described in Figure 12-1.*

This is a familiar process, indeed. It is a feedback control system that enables you to ride a bicycle, catch a ball, learn any motor skill, and keep your body temperature close to constant.

* The correct term is feedback control system, since there are control systems that do not involve feedback. However, technical people tend to omit the descriptor and speak of a control system, feedback implied. Furthermore, many persons refer to the same thing as a feedback system, control implied.

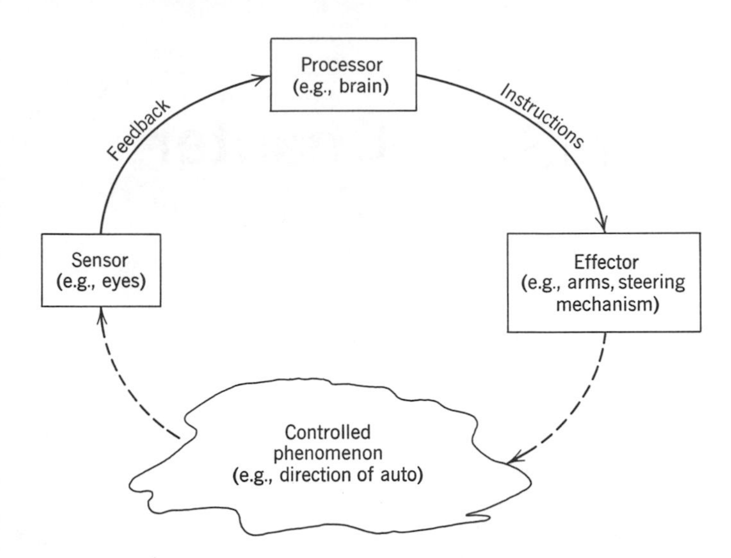

Figure 12-1 *These are the basic components of a feedback control system. A sensor obtains feedback (i.e., a "status report" on what is happening). A processor (decision maker) compares what the sensor reports is actually happening with what is intended (the goal) and decides if and what kind of remedial action is needed. An effector (corrector) executes the corrective action. This cycle of events repeats. You know that as you steer your auto, after you correct your course, you observe the results and if necessary make another correction, and observe the results of it. . . . These activities are usually continuous and simultaneous, not discrete and sequential as described here.*

Even the relatively simple act of reaching to the top corner of this page before turning it involves a rather elaborate control system. The human body contains a remarkable array of such systems, with myriad external and internal sensors, a multilevel decision-making hierarchy, numerous interconnections, and a complex system of glands and muscles serving as effectors. And so it is with all living organisms; recall that even an amoeba can detect and respond to certain changes in its environment.

There are several characteristics of feedback control systems worthy of special mention. One is the closed loop, which is apparent in Figure 12-1. You can understand why the term *closed loop* is often used in referring to such systems. By "closing the loop," effect is coupled with cause, so that the cause-effect relationship is now one of *interdependence*.

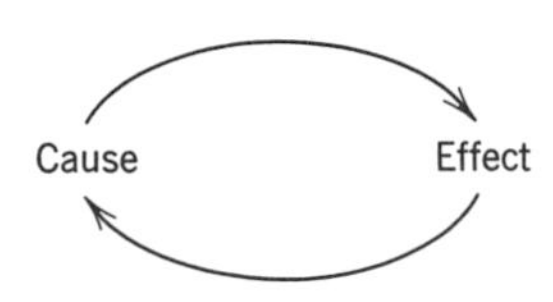

Of course, it is feedback that closes the loop. Feedback is information characterizing the *actual* situation, which the processor compares with the *intended* state of affairs. The discrepancy between intended and actual (the error in other words) becomes the basis of corrective action.

Another feature worth noting is the role of information throughout a feedback control system. The system must be given a desired condition (intended state of affairs, goal, set point). This is information, so is feedback, and so are instructions transmitted to the effector. The processor converts information from one form to another; the effector converts information into action; the sensor converts action into information. No wonder information is sometimes called the lifeblood of a control system.

We should have a feedback system here. If we were speaking face to face, I could get feedback through your facial expressions, questions, and comments to learn if my message is received and understood (my goal). On the basis of the feedback I could decide whether repeating or restating or exemplification are necessary to achieve that goal. Feedback, which one-way communication like the printed page fails to provide, facilitates learning and, in the case of motor skills, is essential. Picture the blindfolded dart thrower; he throws darts indefinitely and shows no improvement as long as he knows nothing about the effectiveness of his efforts. Try it.

Automatic Control

In the control systems mentioned so far, humans predominate. But this is certainly not necessary; man has learned how to create devices that will perform control functions, many of which require no direct intervention by humans. In general, engineers are responsible for designing these *automatic control* systems.

A familiar example of automatic control is the system that maintains the temperature at a comfortable level in modern buildings. The thermostat performs the sensing and processing functions, instructing the furnace (the effector) when to cut in and out in order to keep room temperature (the controlled phenomenon) at the specified level. The human's only role is to prescribe the desired level by setting the thermostat dial.

A classic example of automatic control is the flyball governor applied by James Watt to control the speed of a steam engine (Figure 12-2). This ingenious device made it unnecessary for a man to stand by to adjust the steam input as the load on the engine changed, in order to prevent its speed from fluctuating wildly.

Recall the machine for manufacturing reed switches described

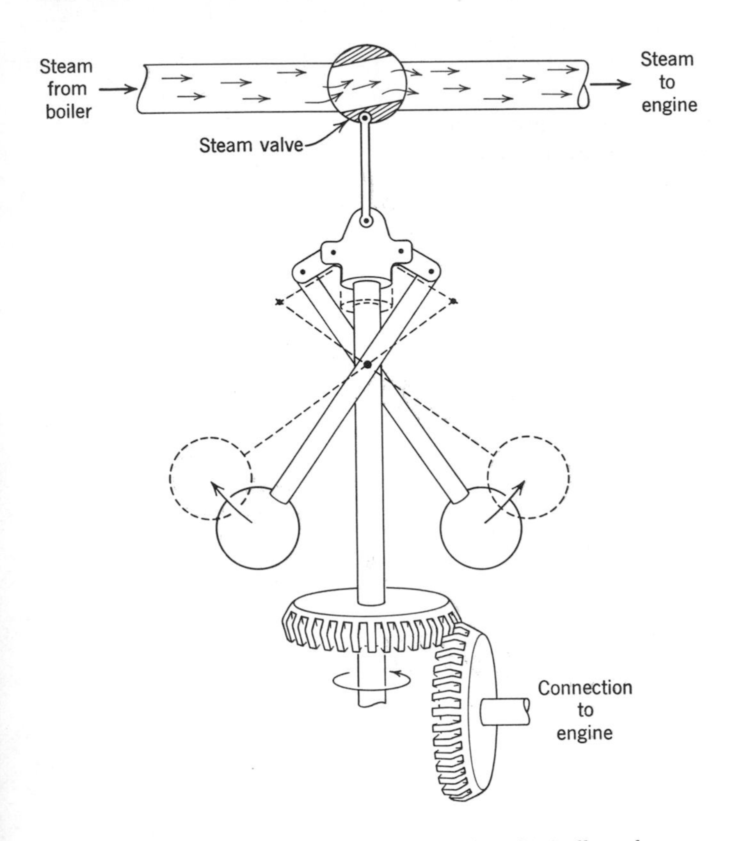

Figure 12-2 *The flyball governor is connected mechanically to the output shaft of a steam engine so that the ball mechanism rotates at the speed of the engine. If the load on the engine decreases, speed will tend to increase which, through centrifugal action, forces the balls outward. Through the linkage, this will proportionately close off the steam supply to the engine. If the engine tends to lose speed, the mechanism increases the steam supply accordingly. Therefore, the flyball governor maintains engine speed at a preset value without human intervention. This invention is significant in several respects. It is remarkable if for no other reason than it was so advanced for its time (the 1780s). Furthermore, it is a classic illustration of the elegant solution. Finally, it is widely recognized as an outstanding example of what engineers can do without the benefit of theory. The mathematical theory of the behavior of this governor did not appear until 1868.*

in Chapter 1. A sophisticated feedback control system is at work. Finished switches pass through a series of instruments that perform critical measurements and tests, not only to eliminate unsatisfactory switches but to enable the production machine to correct the cause of error. If the machine should start producing switches with oversize gaps between reeds, for instance, it automatically detects this and adjusts the gap-setting mechanism appropriately. This machine is a prime example of automation; so are the thermostat system and the ship positioner in Figure 12-3. In each case, the human can walk away, and the machine will continue to perform its intended function.

The same is possible with the automatic washer and the familiar traffic signal, but not for the same reason. These mechanisms are programmed by cams that force the machines to follow a cycle of actions. As long as these machines are functioning properly, they will continue plodding, oblivious of the appropriateness of their actions; the clothes can be clean or filthy or there can be no

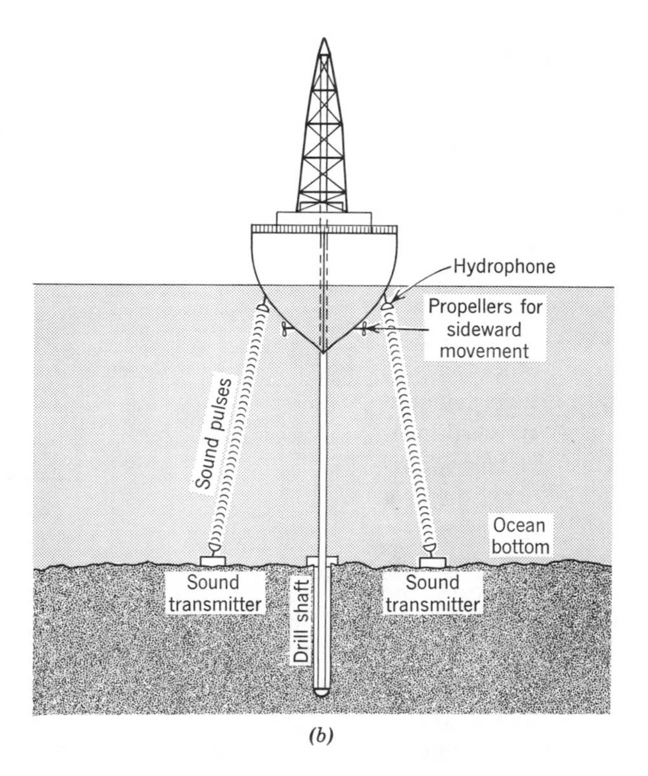

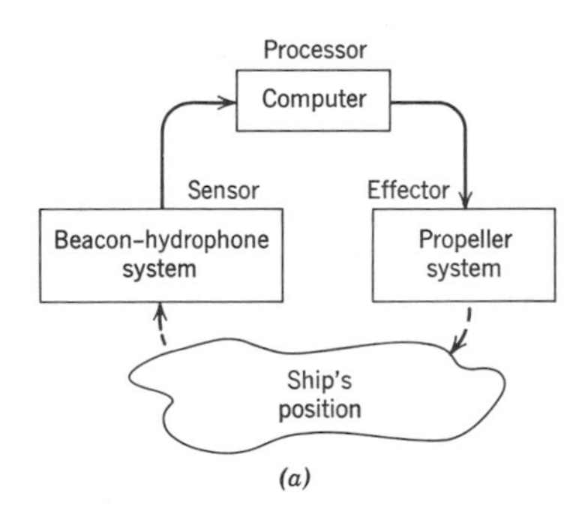

Figure 12-3 *A drillship is an ocean-going vessel equipped with the familiar drilling rig and other special apparatus that enables it to drill into the ocean bottom in deep water. When drilling, such a vessel cannot drift very far from the desired position directly above the drill hole. The automatic control system diagrammed in part* (a) *enables the ship to maintain its position in the face of winds and currents. Beacons* (b) *transmit sound waves from known positions. These are sensed by hydrophones and, from their phase relationship, the computer can calculate drift. On the basis of this, the thrust propellers fore and aft are actuated to return the ship to the desired position.*

clothes in the machine; the traffic can be heavy in one direction or the other, yet the signal cycles on. In contrast, the thermostat, ship positioner, and switch-making machine require no human intervention because they have *feedback* control systems that enable them to adjust their behavior in response to changing conditions. These are often referred to as *self-regulating* or *self-correcting*.

Design of Feedback Control Systems

There are feedback control systems in most engineering creations—automobiles, power-generating stations, dams, elevators, electric coffee-makers, water supply systems, and steel-making furnaces, to name a few. So design of such systems is a rather common activity in engineering. Ordinarily, it all begins with a specific need for control (e.g., a system to maintain airplane cabin pressure within acceptable limits). To synthesize a system that satisfies this need, engineers must know a lot about control systems, in particular such matters as response characteristics, oscillation, sampling, sensitivity, stability, nonlinearity, noise, and a variety of other mysterious-sounding topics.

To give you some notion of what is involved in design of control systems and what engineers must know in order to design them, here is an elaboration on one of the many matters that must be considered: response thresholds (action limits), chosen primarily because no special background is needed to understand what is involved and because it is relevant to control systems wherever they occur.

You have probably noticed that even in a room with a thermostat system, the temperature fluctuates, reaching a low just as the heater cuts in and a high just after it goes off, as graphed in Figure 12-4. This controller has an upper limit and a lower limit, and only when the room temperature reaches one of these limits does the system respond. In effect, the thermostat has been instructed not to bother the furnace until the temperature changes enough to make startup worthwhile. The thermostat obliges; it "calls" to the furnace for heat only when room temperature falls 3 degrees below the setting on its dial. When heat is called for, room temperature is allowed to reach 3 degrees above the set value before the furnace is signaled to quit.

Response thresholds of a control system bound an *insensitive zone*. Within this zone the controlled phenomenon can fluctuate without the effector knowing or caring. In many cases it is

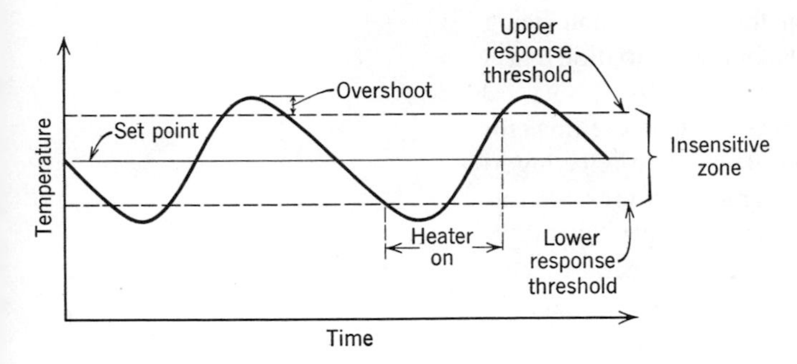

Figure 12-4

infeasible to do otherwise. Take the thermostat system; without an insensitive zone surrounding the set temperature, the controller would go berserk—so would the furnace—going on and off in rapid succession as the room temperature oscillated fractions of a degree above and below the set value.

Here is another use of response thresholds. It is possible to attach a tiny sensor to a critically ill patient to pick up his heart beat. These pulses are transmitted to a monitoring unit at the nurse's desk (Figure 12-5). That unit sounds an alarm if the patient's pulse rate goes above or below preset limits (right). It

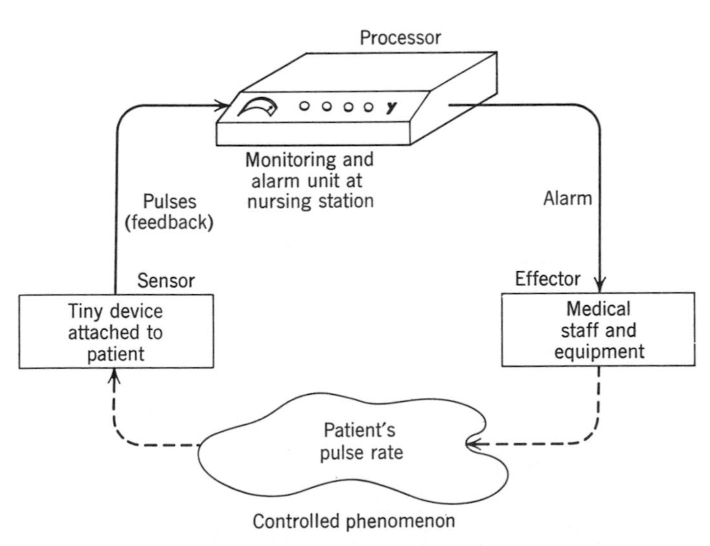

Figure 12-5 *A feedback control system for maintaining a patient's pulse rate within acceptable limits.*

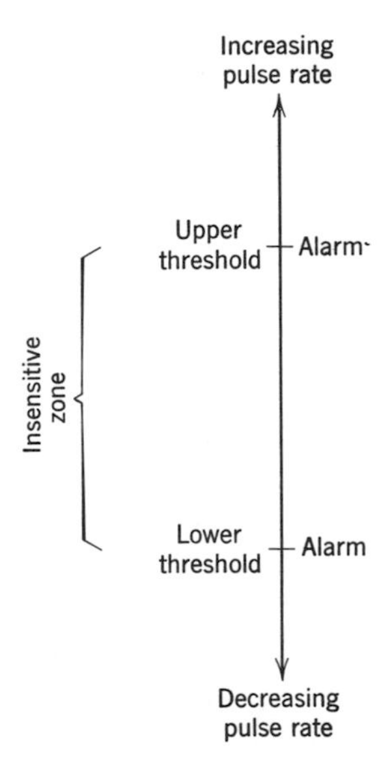

makes sense to use such limits when the phenomenon being controlled is subject to a certain amount of natural variation that is no cause for alarm.

The manner in which an engineer arrives at the best settings for such limits is an excellent study in optimization. In referring to Figure 12-6, picture an engineer designing a system for controlling water level behind a dam. It is infeasible to control that level to an exact value; a certain insensitive zone is a natural for this situation; but how wide should that zone be? As the zone is widened, there is an increase in the penalties and inconveniences associated with fluctuations in lake level (curve *a,* Figure 12-6). As the zone is reduced, more expensive control equipment is necessary, and the cost of operating and maintaining the spillway gates increases because they must respond more frequently (curve *b*). Because of these conflicting criteria, there is an optimum zone width, determined by the minimum point of curve *c*.

Incidentally, if you try to determine the optimum thresholds for

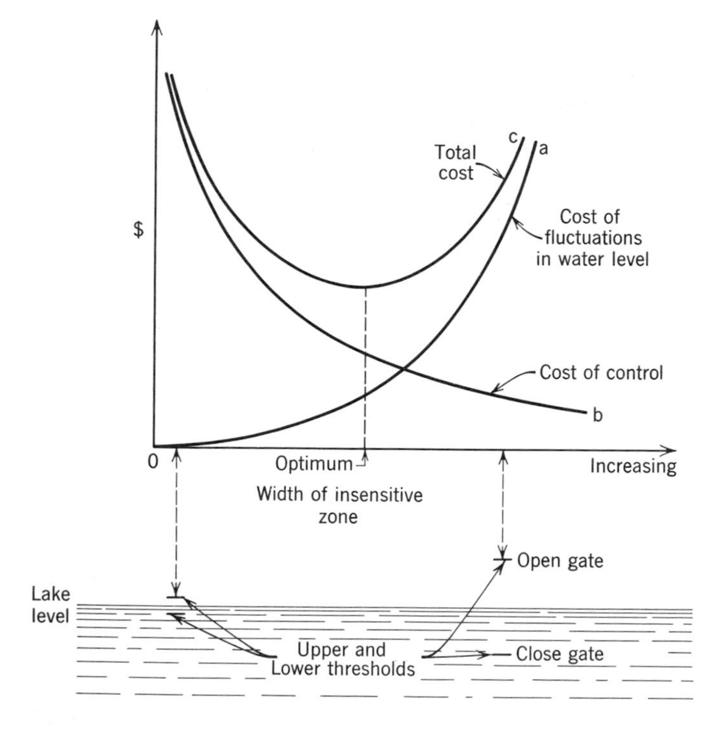

Figure 12-6 *An illustration of optimization in design, in this instance, the design of a system for controlling water level behind a dam.*

the pulse-rate monitor described by Figure 12-5, you again run headlong into the life-versus-dollars dilemma. The trade-off is between staffing cost and risk to the patient. As you tighten the insensitive zone, you increase the number of false alarms; as you widen it, you increase the probability that a patient requiring emergency attention will not be discovered until too late.

The setting of thresholds is only one of many problems with which the designer of a control system must cope. Probably the thorniest one is oscillation—the tendency of the controlled phenomenon to vibrate about the desired value. There are several types of oscillation. One is continuous, which is nicely illustrated by the temperature control situation in a classroom that I use. There is no thermostat; temperature is instructor-controlled and, usually, it goes like this. The first instructor to use the room that day finds it cool when he arrives. He immediately turns the radiator valve on full and forgets it. By the time the next instructor enters the room it is unbearably hot, so his first act is to shut off the heat. Of course, he doesn't turn it on again before he leaves. And so, throughout the day, the temperature oscillates between uncomfortable extremes. Continuous oscillation like that demonstrated here is commonly referred to as *hunting*.

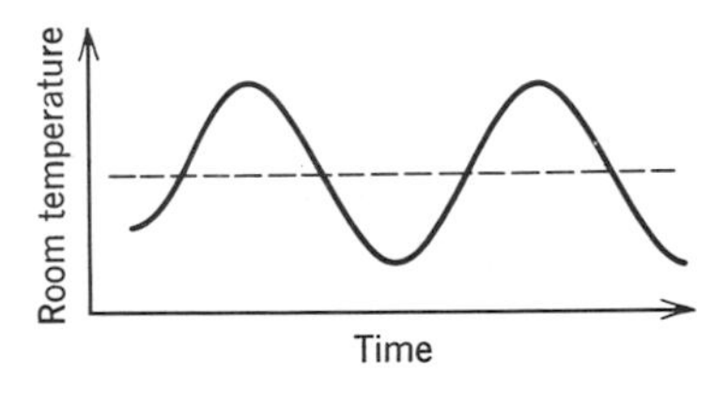

You don't think that the flyball governor returns engine speed immediately and exactly to the selected speed, do you? When speed changes, the governor brings it back toward the selected valve but, because of momentum, it overshoots. So it must swing in the opposite direction and, again, it overshoots, and so on. But, because of friction, this oscillation subsides—dampens out, as engineers say. In some cases this type of oscillation is inconsequential; in others it is intolerable and complicates the control problem immensely.

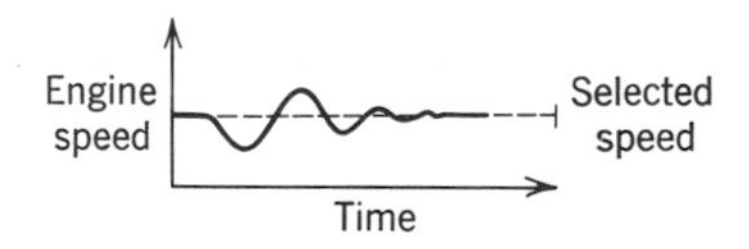

There are cases in which oscillation, once triggered, continues to grow. An unstable situation like this could be disastrous. It will probably happen to you when you try to walk on a railroad track. In that case it will probably be of no great consequence, but a hovering VTOL that develops this type of oscillation is in serious trouble.

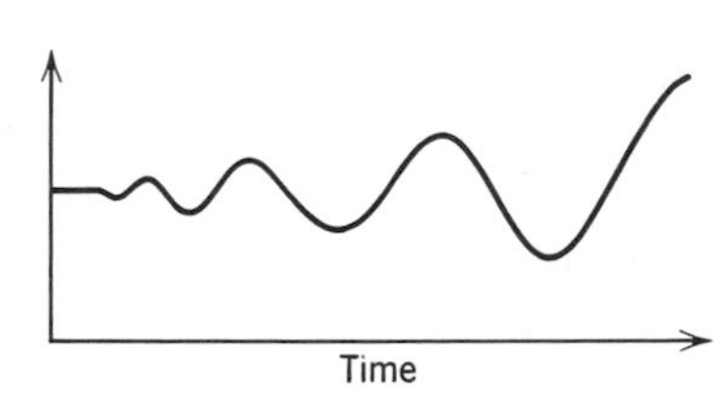

Ordinarily, oscillation of a control system is the result of delayed response (i.e. delay between the time a change occurs in the controlled phenomenon and the time the system responds). This *lag* causes overshooting—back and forth over the desired level. Lag may arise anywhere in a control system; for example, the feedback may be delayed or the effector may be slow in responding. Certainly a furnace does not produce heat the instant

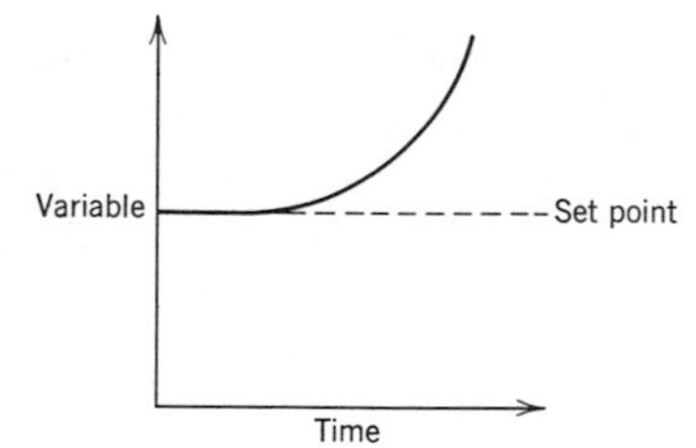

Figure 12-7 *Speaking of instability! In contrast to oscillation, there is another form of malady in which the system, once it goes out of control for whatever reason, continues in that direction at an accelerating rate until something "gives." The cupcake machine obviously has this problem.*

it is triggered, nor does it cease doing so as the power is cut off. Such sluggishness and the resulting oscillation is often apparent in economic and political systems. All control systems suffer from some lag and, therefore, are prone to oscillate to *some* extent. For the engineer it's a matter of keeping that oscillation at an optimum level.

Modeling of Control Systems

Picture yourself designing a controller for a sprinkler system, the overhead type found in many larger buildings for fire suppression. This controller is to open the water main when room temperature reaches 140 degrees. You have three alternative designs in mind, similar in cost but perhaps not in performance. You must determine which design is best with respect to accuracy, consistency, and speed of response.

Naturally, you are going to use the quickest and simplest means of evaluating them. Why not have a breadboard version (the term for a simple, crude working model) constructed for each of your designs and experiment? Perfectly logical; you could move a hot plate toward the sensor of each device and record data such as temperature and time of response. This could be repeated numerous times for each design in order to evaluate consistency of behavior, which is at least as important as any other performance characteristic. (An inconsistent controller frequently and erratically triggers above and below the desired temperature.)

The point of this story is that your alternative designs can be evaluated satisfactorily by *experimentation*. You are systematically altering the "input" (stimulus) to each model in order to measure certain characteristics of the output (response). All in all, it's a relatively simple process.

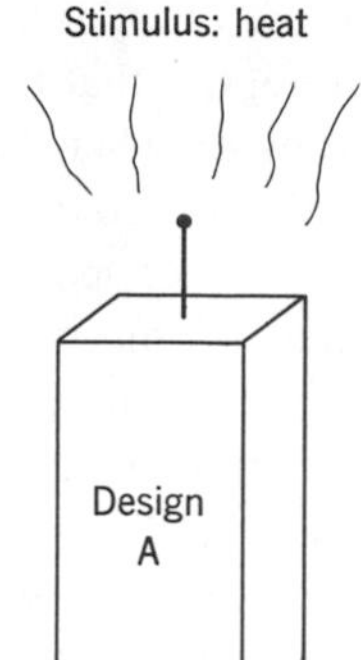

However, it's a different story for the engineer designing the control system for maintaining a drillship in position (Figure 12-3). No simple experiments are feasible in this instance, and so he uses mathematical models in a manner that closely parallels your use of experimentation with alternative sprinkler system controllers. He assumes different out-of-position conditions, then relies on his equations to predict behavior of alternative designs, such as their speed of response and tendency to oscillate.

So design of a control system calls for some familiar skills (e.g., *definition* of the control problem, *search* for solutions, *modeling* to predict performance) with which the engineer synthesizes a device that will best perform the control function given.

Your Knowledge of Control Systems

Indeed, engineers must know a lot about the behavior and synthesis of control systems, and there is much to know, which will be readily apparent if, when you are in the library, you note the number of books on the subject and skim a few. Books, courses, and sequences of courses are available.

This knowledge is required primarily for the design of complex control devices, but it has secondary benefits. For instance, knowing the basics of such systems enables you to appreciate better what humans might do as they use their technology to tamper with nature. Nature employs a wide variety of control systems to maintain stability (of populations and the environment, for instance). Undisturbed, nature keeps things pretty well under control. But man, by crippling or eliminating a natural control mechanism, can indirectly cause severe if not catastrophic effects. There are many instances in which populations of fish, rodents, insects, and other forms of wildlife have gotten out of control because man interfered with the control mechanisms nature provided (e.g., predators, food supplies, or the environment). These cases of man-caused instability have been less than catastrophic, but that may not always be so.

Furthermore, what you have gained from this introduction will enable you to recognize some of the maladies that political, social, economic, and other institutions suffer as a result of their flimsy feedback mechanisms. An institution (a city government, for instance) is a large, complex, clumsy organism that, like its animal counterpart, requires numerous feedback systems. But, in general, the feedback mechanisms of institutions are crude at best, to the extent that armed only with an elementary understanding of feedback principles, you can probably identify flaws and potential improvements.

Obviously we have returned to the recurring theme concerning the relevance of subjects covered in an engineering curriculum to nontechnical problem areas. In the case of feedback control systems, the readings cited at the end of this book reinforce this point. Sample those readings not only because control systems are significant in the creations and affairs of men; you are likely to find it a fascinating subject.

EXERCISES

1. To help you recognize the pervasiveness of feedback systems, list examples of such systems spanning engineering, biological, political, business, learning, social, and economic phenomena. Variety as well as number will be valued.

2. What do you think keeps the blades of a windmill facing into the wind? You probably never gave the matter any thought. Please do. In fact, do a little investigating in the library and report your findings in a minipaper.

3. Oscillation (hunting) is hardly confined to engineering contrivances. Surely you can identify examples in economic, social, and biological systems. For each one, explain what could be done to reduce if not eliminate the oscillations.

4. In this chapter you read: "For the engineer it's a matter of keeping that oscillation at an *optimum* level." Why optimum rather than minimum?

5. Why does a business enterprise need a variety of internally and externally oriented feedback systems? Identify some specific kinds of feedback systems that an enterprise *better* have.

6. You and your instructor are each a principal part of a feedback system in this course and in any other course.

(a) **Diagram the system in which each of you makes the decisions. Label the major components. Your diagrams might resemble Figure 12-1.**
(b) **Explain what form each major component takes in your learning situation (e.g., your instructor "senses" your progress through recitation).**
(c) **Identify some of the important flaws in these systems.**

7. List the feedback control systems to be found in an automobile or airplane. Some you already know exist; others you can learn about through reading and questioning.

8. Now the term pollution control should have added meaning. Following the pattern set by Figure 12-1, diagram your vision of an overall pollution control system for the region in which you live. Assume that the "processor" is a hypothetical regional pollution control agency.

DIGITAL COMPUTERS Chapter 13

To receive the full benefit of this chapter, you do not have to know the details of computer operation. The emphasis here is on applications. However, it will be helpful if you know a few general facts about these machines, beginning with Figures 13-1 and 13-2.

When a computer is used, the sequence of events is generally as described on pages 222 and 223.

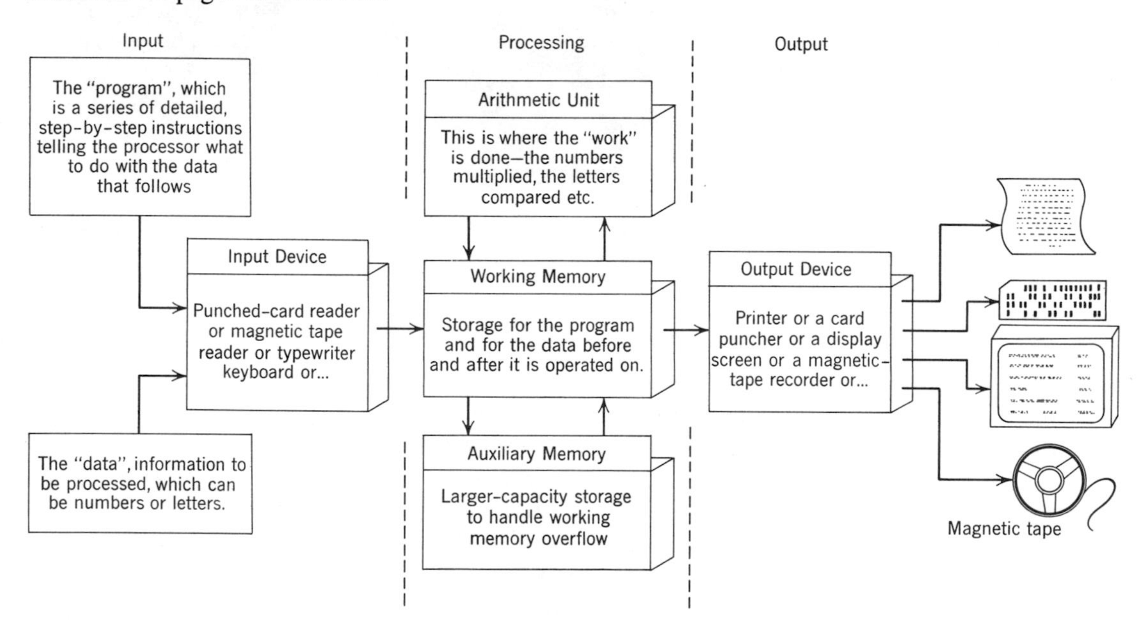

Figure 13-1 *A simplified overall view of a computer system. The processor's "working memory" serves much the same purpose your mind serves in memorizing computational procedures* and *paper serves when you are doing arithmetic by the pencil and paper method. Although this memory is incredibly fast, it has a relatively limited capacity and, furthermore, it is "cleared" after each job is complete. So, when more capacity is needed or when the user wishes to preserve the memory's contents, an auxiliary storage device like magnetic tape or a magnetic disk augments the working memory. Auxiliary storages are generally slower, but they do have virtually unlimited capacities for letters and numbers–billions of them.*

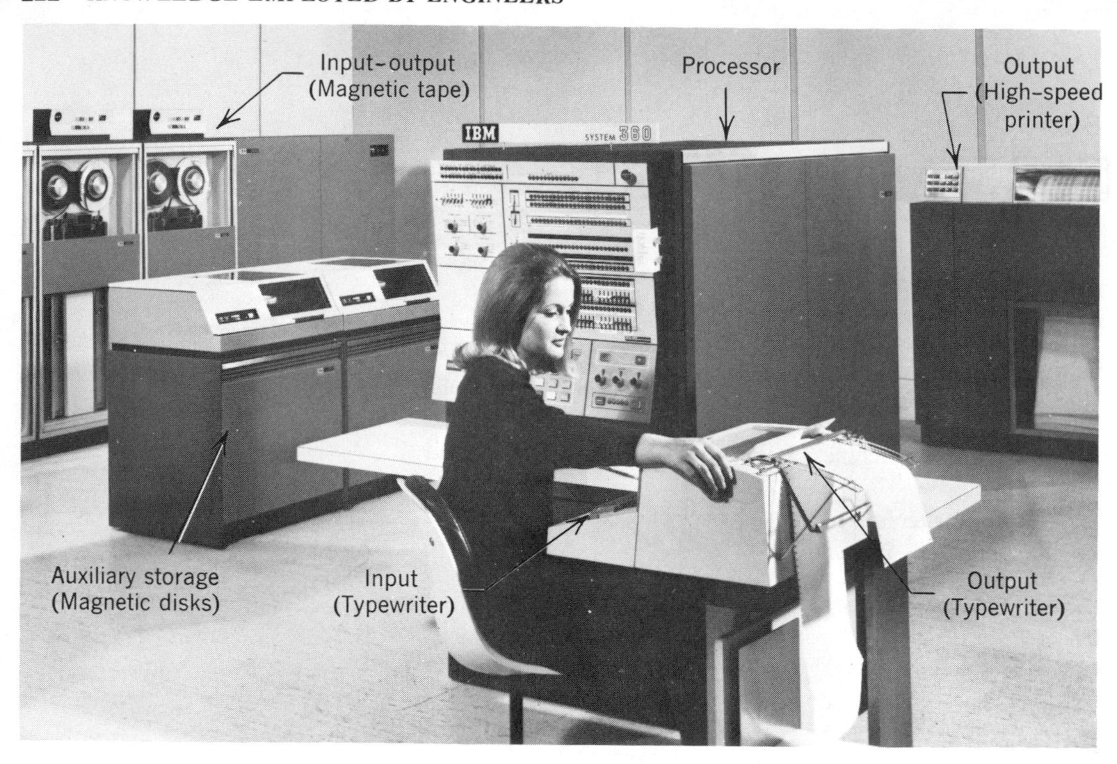

Figure 13-2 *Examples of the computer equipment (computer people say "hardware") identified in Figure 13-1. Not shown is a card reader-punch (a machine that senses the information punched in cards or punches information into them), a paper-tape reader-punch, or a cathode-ray-tube terminal (computer buffs say CRT). A CRT terminal (Figure 13-5) is a close relative to a television receiver in appearance and operation.*

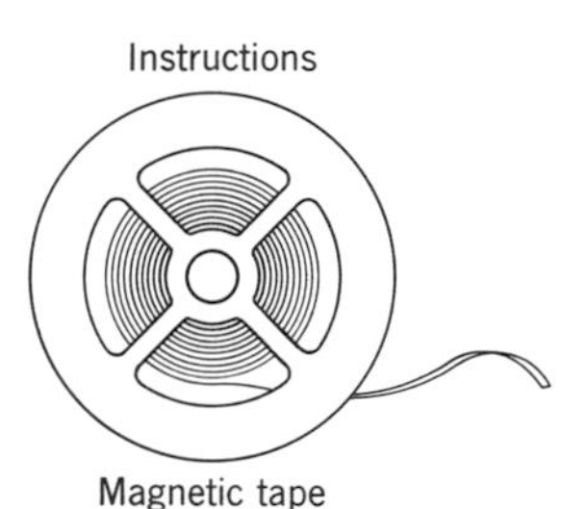

1. **A program of instructions is prepared (or, since programs for many purposes are already prepared, it may be simply a matter of selecting one). These instructions tell the computer what to do with the numbers and letters subsequently given to it, step by step (e.g., "take the number stored in memory location A and add it to the number in memory location B, place the sum in memory location C"). The program even includes details such as where to print the results on the paper. Incidentally, programs are written in terms of "unknowns," like the X's and Y's of algebra. However, in the program you can use words for your unknowns, as illustrated by this excerpt from a program for computing the take-home wage**

of an hourly employee.
WAGE = (HOURS × RATE) − HOSP − INSUR − TAX
Each week, when employee earnings are to be calculated, the same program is entered into the computer and stored in the working memory.

2. **Next, the data to be processed are entered into the computer and stored in another part of its memory. In the payroll example, the data consists of the identification number, hourly pay, number of hours worked, and so forth, for every employee. If the number of employees is large, not all of these data will fit into the computer's memory at one time; they will be entered in reasonable-sized batches.**
3. **Now that the data *and* instructions are in the machine's memory, the processor begins operating on that data according to those instructions. In the payroll example, the same calculations must be performed for every employee, so the processor simply recycles and repeats the same series of instructions for each employee.**
4. **Eventually the results of the computer's operations are communicated to humans by printing them on paper or by other means. In the payroll example, the machine is instructed to print and punch the results on cards, which become paychecks.**

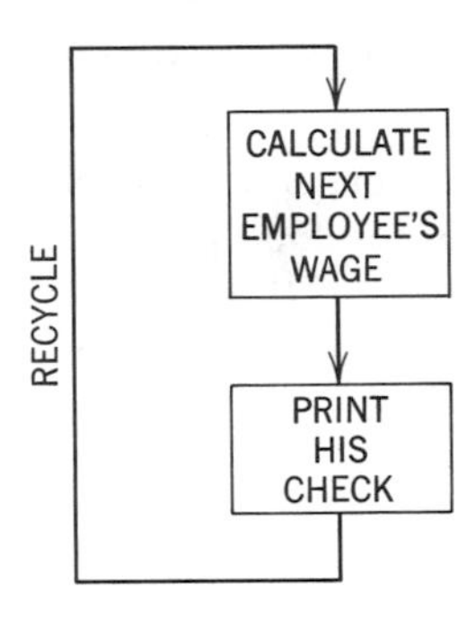

A digital computer can accept, transfer, and manipulate numbers *and* letters, including letters combined into words and groups of words. It is capable of performing all arithmetic operations. One especially significant capability of a digital computer that you may not know about is its ability to compare numbers *or* letters and follow different series of instructions, depending on the results of the comparison. Suppose a computer is being used to keep driver and auto records for a state. When an application is received for transfer of ownership, part of the computer's task is to check if the auto has been reported stolen. It does so by comparing the auto serial number (e.g., AXB1372594) with each serial number in a list of serial numbers of cars reported stolen, stored in memory. When it compares AXB1372594 with a number on this list, if the two do not match, the computer goes on to the next number (Figure 13-3). As long as it finds no match, it continues through the list. If the two match, the computer follows an alternate series of instructions that cause it to print out

AXB1372594 REPORTED STOLEN JAN 6, 1976 BY JOHN REES 721 CENTER STREET BIRDSVILLE MICH

This important feature, called *branching*, gives the computer more flexibility than you may have given it credit for. There may be thousands of such branches in a long program, so that the computer rarely repeats the same sequence of instructions. Among other things, the computer's branching capability enables it to handle exceptions (e.g., an employee who worked more than the usual 40 hours) (Figure 13-4). You may recall seeing such a diagram earlier (page 161) explaining how a computer can execute the Monte Carlo process.

Of course, there are many variations in computer equipment and procedures for using it but, on the basis of this background, you can develop a general appreciation of what digital computers do.

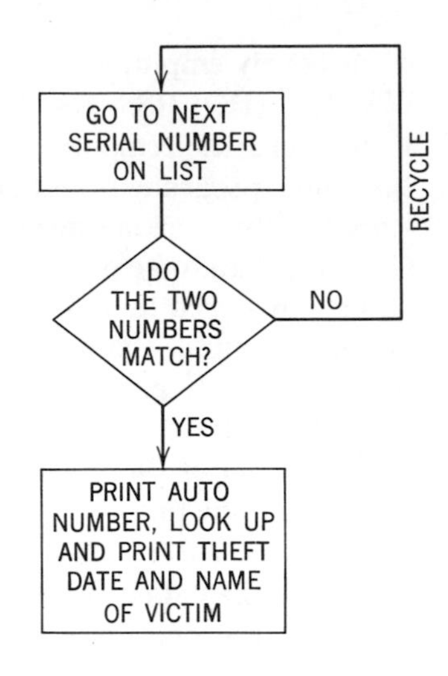

Figure 13-3

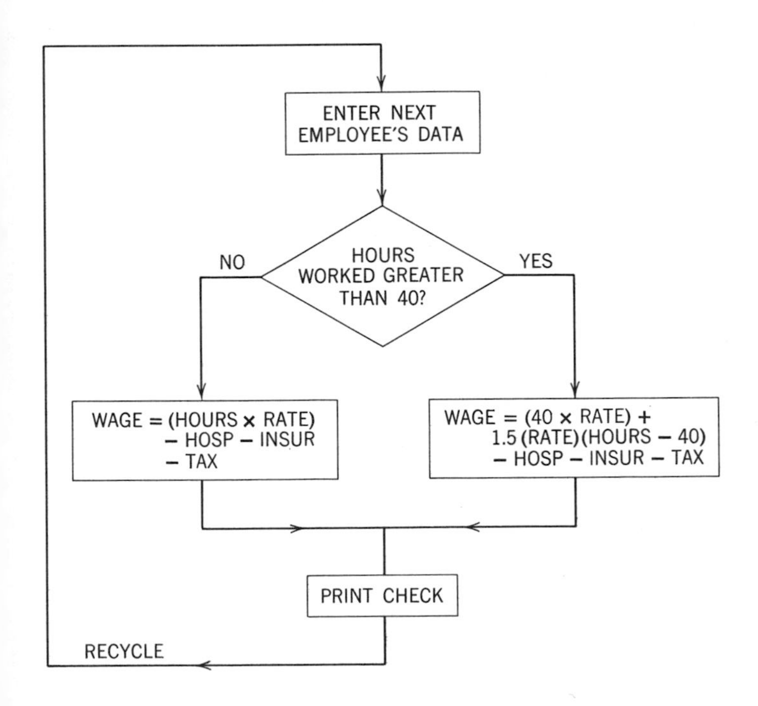

Figure 13-4

AN OVERVIEW OF THE ROLE OF COMPUTERS IN ENGINEERING

Ability to program a computer (i.e., to prepare a series of instructions that enables it to successfully complete a computational task) is a skill that many engineers find useful, to the extent that most students learn to do so early in their engineering curriculum. But engineers should also be generally familiar with the broad applicability of computers. This you can acquire from what follows, while at the same time you are observing a sample of the kinds of knowledge (versus skills and attitudes) that an engineer should possess.

This survey of the usefulness of computers to engineers focuses first on the ways engineers use a computer as a tool in arriving *at* solutions to problems, in the same sense that the hand calculator and transit are tools. Later you will learn how computers are utilized *in* solutions (e.g., as part of a spacecraft guidance system).

The Digital Computer as A Problem Solving Aid

The computer has certainly affected the practice of engineering. It is fast becoming an indispensable design tool, aiding engineers in such ways as these.

Literature Search

An engineer is designing a comprehensive computer system for a university. This system is expected to process grades, prepare schedules, handle accounts, maintain student records, do some instructing, and perform a multitude of other useful functions. The engineer surmises that systems to handle various parts of this problem have been developed by individuals at other institutions and companies, but how does he learn who the people are and what they have developed? He could easily spend several weeks tracking down reports of the work done on this sort of thing elsewhere, through library searches, reading, talking, corresponding, telephoning, and traveling. And, after all this effort, he would have missed some worthwhile systems that others have worked out simply because there are so many scattered references and because an exhaustive search would be too expensive and time-consuming.

This is the way it is for most engineering problems. The literature search tasks are staggering. Ordinarily the engineer conducts a reasonable search and then proceeds to solve the problem at hand, accepting the risk that he is duplicating the work of someone who has already solved part or all of the same problem. Under the circumstances, it is uneconomical for him to do otherwise.

But the computer is coming to the rescue, enabling the engineer to conduct his search as follows. He lists key words, the kind of words he would keep in mind if he were conducting a search in the library. In this case he would have two lists (margin). He submits these to his company's computer center to be used for search of its files on computer applications. These files consist of magnetic tapes containing abstracts that describe every computer application made public. The computer goes through these tapes and, for every abstract that contains at least one word from list 1 *and* at least one word from list 2, it prints the abstract and the information the engineer needs to get the details. Within the half hour he can obtain a more comprehensive collection of references than he could possibly afford by other methods. (Here the engineer is employing a computer as an aid in design. In this project a computer will also be part of the system he is designing.)

Keyword List 1	*Keyword List 2*
College	Admissions
School	Billing
University	File keeping
Student	Inventory
	Record keeping
	Registration
	Scheduling

Such search systems are not in widespread use, but you can see why they are under development. They minimize the costly duplication of research and problem-solving efforts and eliminate the need for arduously sifting through "mountains" of books and technical reports. They do the job quickly and exhaustively.

Data Reduction

Engineers often have large volumes of data to be reduced to useful form, for example, hundreds of measurements from an experiment. Calculating averages and measures of variability, curve fitting, statistical tests, and the like, are usually time-consuming and tedious if done by hand. The computer is a natural for such chores.

Mathematical Operations

Most common mathematical operations can be executed by the computer, and thank goodness. Some of them, like solving simultaneous equations of many unknowns, take hours or days to do by hand. Unlike the equations ordinarily found in mathematics books, those encountered in practice are often messy to solve. The $y = ax^b$ of the textbook is likely to be $y = 1.878x^{2.3}$ in practice. As a result, computation can become very time-consuming. Try finding the value of x that balances the equation $x = 1.31e^{0.27x}$ to within 1 percent. This will take some time by pencil-and-paper methods. You can see why the computer is so important to engineers; without it such chores are a significant time drain and, moreover, a big bore.

Iterative Optimization

Recall from page 191 that iterative optimization can be time-consuming if done by hand. In fact, optimization by formal iterative methods was not often attempted, even considered, before computers became available. Now engineers can take full advantage of this powerful technique.

Simulation

A fast-spreading technique is digital simulation by computer. Since simulation *is* experimentation on a model, and since the whole operation can be computerized, the engineer can, in this sense, conduct experiments on a computer. (You know why engineers turn to the computer to carry out digital simulations if you have ever done one by hand!)

The "Computer Draftsman"

Auxiliary equipment now enables an engineer to communicate graphically with a computer. This remarkable and significant development is just beginning to affect the practice of engineering but, within a few years, the impact of *computer graphics* will be dramatic. You can see from Figures 13-5 to 13-8 that this capability has enormous potential.

In this sampling of the ways that engineers use computers, there are applications that are close to commonplace and others that are just coming over the horizon. While gaining insight into the usefulness of computers in engineering, you are getting a good idea of what they offer to architects, businessmen, physicians, and other problem solvers.

Figure 13-5*a* *This CRT enables a computer to display letters and numbers. The user communicates to a computer using a typewriter keyboard or the light pen he holds. This physician can indicate which of the tests listed on the screen he wants performed on a patient by pointing with the pen.*

Figure 13-5*b* *This combination of computer, CRT, man, and light pen can accomplish remarkable things. Man can "draw" on the face of the tube with the pen, and the computer can remember what is drawn and reproduce it on the screen whenever it is instructed to do so. Even better, the computer can manipulate these drawings. For example, it can show the object from different perspectives, rotate it, or enlarge it, as instructed by the pen and the keyboard (Figure 13-6). In this particular view, the operator has pressed the "line deletion" button, which enables him to point to any line and have it erased from the screen and the computer's memory.*

Figure 13-6 *These before and after photographs of the screen pictured in Figure 13-5b illustrate some of this system's capabilities. When the engineer finishes his work and wants to make a permanent record of what appears on the screen, he does so by instructing the system to store it, in which case it can be called back to the screen at any time. He can also have the computer instruct a plotting machine, which will prepare an inked drawing, or he can obtain a permanent copy by photographic means.*

Utilization of Computers IN Solutions

In addition to aiding the designer as he arrives at a solution, a computer often becomes a part *of* his solution, for one or more of the following reasons.

1. **The solution requires a means of *storing and retrieving* information, and a computer is the most effective alternative.**

Give the computer these:

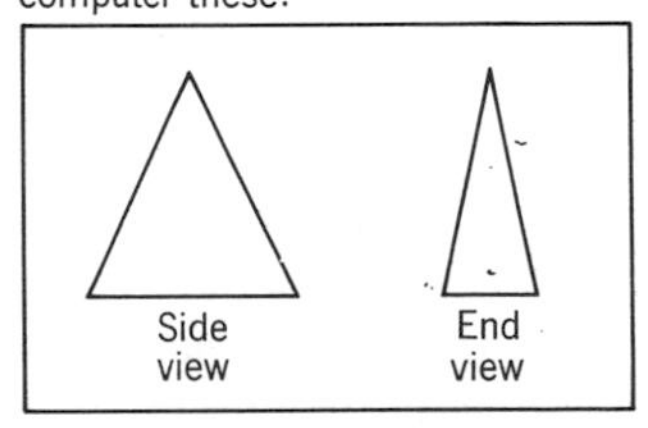

and it will give you

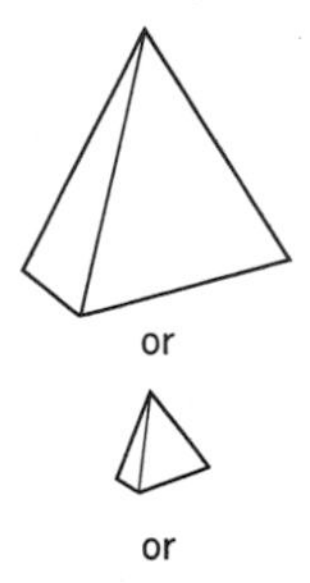

or

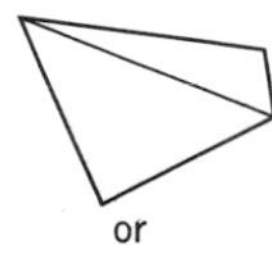

or

⋮

2. **The solution requires a means of *processing* information, and a computer is the most effective alternative.**
3. **The solution requires a means of handling information quickly, and *only* a computer is fast enough.**
4. **The solution requires a means of keeping track of many concurrently changing, interacting events or variables, and *only* a computer is capable of it.**

Elaboration follows, but first, note the generality and usefulness of this classification of computer applications; it is the pattern of applications in education, business, government—everywhere.

1. **Information storage and retrieval.** The preservation of knowledge so that it can be retrieved without unreasonable time and expense is a problem in many fields of endeavor. The story is the same in medicine, law, business, education, and government, as well as in engineering. Take medicine; with the thousands of periodicals, books, and reports that fill libraries, how does a physician learn quickly whether anyone anywhere has successfully treated a case of Ozarkmumblitus? Problem solvers in other fields have related information search problems; brains, books, and file drawers—familiar means of storing information—no longer suffice. But the computer has a memory that is remarkably reliable, large, and fast. It offers real hope.

Here is a sample application. Representatives of a state government requested a consulting firm to develop a more effective system for storing and searching the state's legal information, which includes all statutes and all court decisions. The heart of the solution to this problem is a computer. With this system lawyers, judges, and legislators can search the state's statutes in order to isolate all laws pertaining to a given subject. If a legislator wishes to know what laws have something to say about the education of handicapped children, he prepares keyword lists and submits them to the computer on punched cards. The computer subsequently prints the numbers of all statutes that contain the words *education, handicapped,* and *children* (or synonyms of these words). In fact, it will print the relevant statutes themselves if so instructed.

This class of applications, illustrated by these medical and legal storage-retrieval systems, is referred to as the computer's library function, since it concerns general knowledge, much of which is available in libraries. But man has a second major type of information storage-retrieval problem. It involves private information (e.g., an insurance company's files on its policyholders)

Figure 13-7 *Given a series of numbers that locates the corners of an object as* X-Y-Z *coordinates, this computer system will prepare television pictures of that object at any perspective and to any scale. Since it can change views up to 30 times a second, it can give the illusion of movement. Views like those shown here can be made to change "continuously" in response to movement of typical aircraft controls, thus allowing a pilot to practice landings on this carrier–in color, at that. Obviously this system is useful for simulation.*

Figure 13-8 *The computer system referred to in Figure 13-7 can do "visualization wonders" for engineers and their clients. Given a minimum of information in terms of coordinates, the computer can "take you for a ride" along a stretch of highway that, at the moment, exists only as a concept. This photo was taken directly from a CRT screen during such a ride. The same can be done for a building–it can be viewed from any perspective as it would appear to an external observer, and views of the external world can be displayed as they would appear to an observer within the building. These feats give you some idea of the exciting things that are coming in the field of computer graphics.*

now typically stored in the office files of corporations, government agencies, schools, hospitals, and numerous other institutions. Computers are well suited for this type of storage-retrieval task also. It is called the filekeeping function.

A hospital has a staggering filekeeping problem. Much time is spent in recording, filing, exchanging, and looking up information. The record-keeping task—ugh! Then there are the thousands of square feet consumed by patient records. A team of engineers called on to solve a hospital's "information problem" proposed the computer system summarized by Figure 13-9. The benefits are impressive. This system keeps an up-to-the-minute record for each patient and answers inquiries within a few seconds. A sampling of the services it provides: checks prescriptions for inconsistencies and errors, reminds nurses when it is time to administer medication; and tells the kitchen how many pounds of each type of food on the menu will be needed for the next meal.

You might wonder about the possibility of objectionable delays, what with 30 typewriter terminals located around the

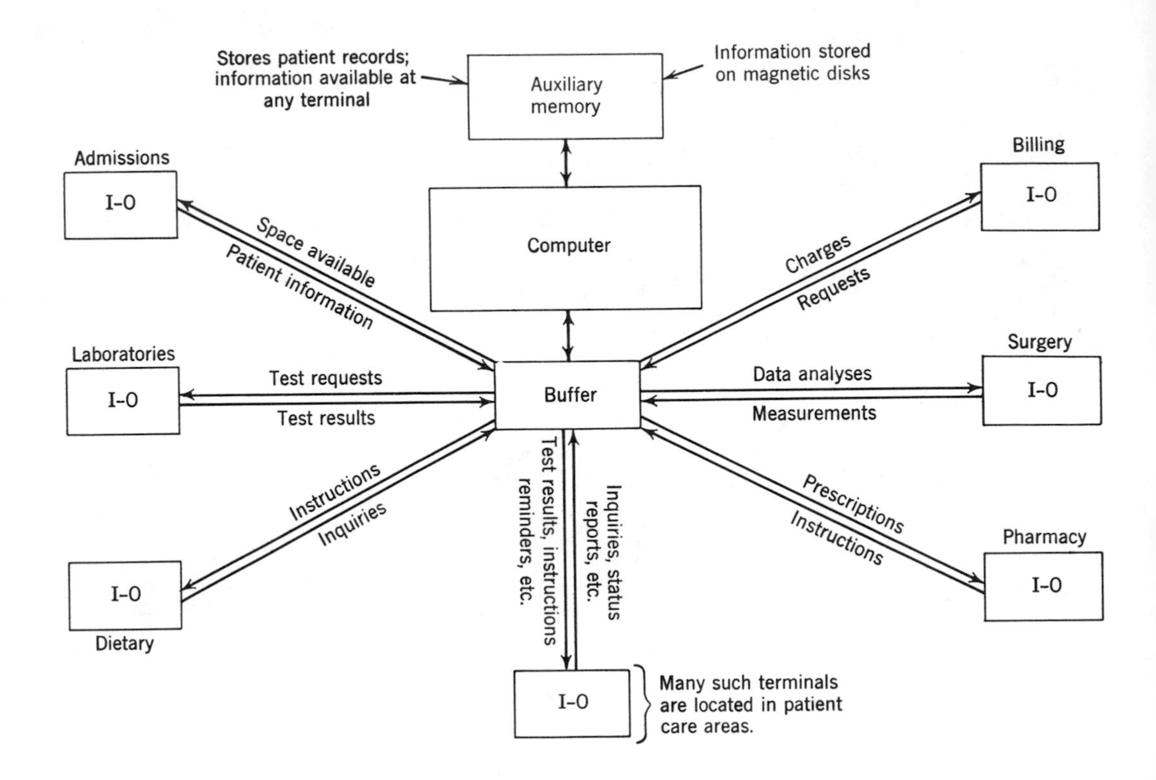

Figure 13-9 *A "computerized" master record-keeping system for a hospital. From any of the terminals located throughout the hospital, an authorized person can add information to a patient's file or request information from it. Examples of data this system stores, processes, and transfers are indicated by the arrows. Equipment at a typical input-output station is pictured by the inset. This computer can also serve as the heart of a patient-monitoring system that alerts hospital personnel when there is a change in the vital life signs of a critically ill person.*

hospital—all wired into the same central computer. Delays of more than a second or two are very unlikely in view of the computer's speed and of the manner in which the system operates. It scans all terminals many times a second in round-robin fashion. When it encounters a terminal at which there is a customer (call him user A), the computer devotes 1/20 second to his request (which may be to file, look up, or process information). If it finishes user A's job in that interval, as is likely, fine. If it doesn't, it puts his job on ice, so to speak, and it scans to see whether there are other active terminals. Scanning takes negligible time. If there are no other terminals at which someone has a task, the machine comes back to user A and gives him another 1/20 second. Then it scans again. But suppose that this time it finds another active terminal, at which user B has work for it. It gives user B 1/20 second, scans and, finding no new active terminals, it returns to user A. And so it goes. Even if most of the 30 terminals are active simultaneously, which is very unlikely, most users' tasks will be completed quickly enough to give each the impression that he has exclusive use of the computer.

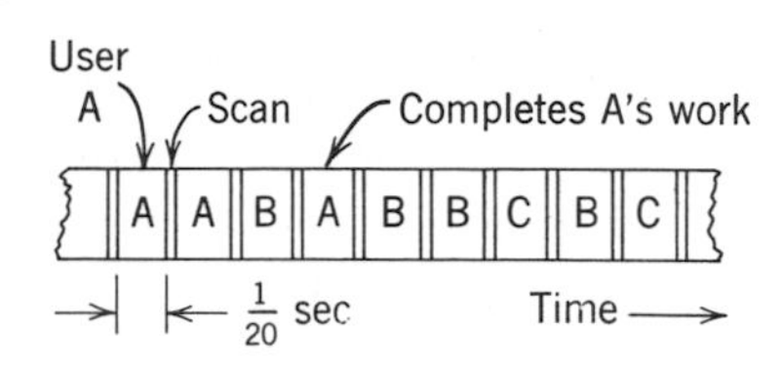

This mode of operation, called *time sharing,* is a relatively recent development in the computer field. You will hear a lot about it and surely feel its impact. To believe that a system like this exists requires that you have a feel for the quickness of the computer. Actually, it is serving its customers sequentially, but it switches so rapidly and completes its work so quickly that it appears to be serving them simultaneously.

The communication buffer between man and computer (Figure 13-9) is an ingenious device without which time sharing would be impractical. User *A* spent 7 seconds typing his request for patient information. Fortunately, during that period, his typewriter was not transmitting directly to the computer. His message went to the buffer, where it was stored until user *A* pressed the end-of-message key. Then, when the computer scanned to user *A,* the buffer fired his message into the computer in a few milliseconds. This frees the very fast and expensive computer from the relatively slow process of typing. The buffer, then, serves as a "time compressor" for incoming messages. It serves also as a "time expander" in the reverse direction, so that the 0.2-millisecond message it receives from the computer is fed to the typewriter at a speed the latter can follow. Clever.

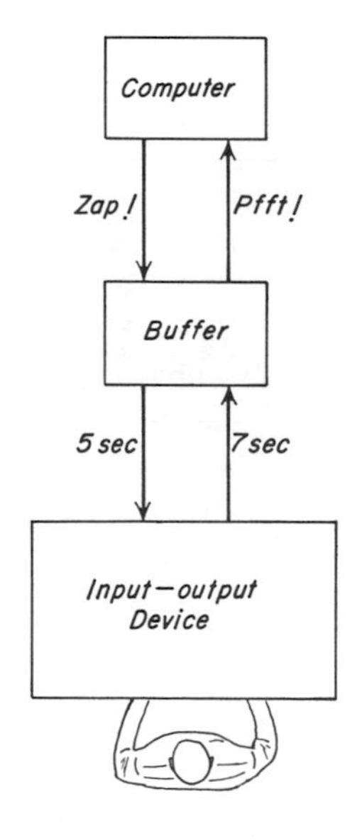

This description of a computerized master information system for a hospital is fiction; some parts of it are operating at some hospitals, but nothing like this comprehensive "manager of medical information'' currently exists. However, you have had a glimpse of what is coming not only for hospitals but for insurance companies, motor vehicle bureaus, local governments, mail-order houses, school systems, and so many other enterprises in which the record-keeping problems are compounding.

2. **When the computer is the most effective method of processing the information at hand.** In the preceding applications information storage and retrieval were primary, and processing of information was incidental. Here the reverse is true. Probably most computer applications you know of are of this type, which is understandable; this *is* the most frequent type of application.

A computer is likely to be the cheapest way of doing the job whenever large numbers of repetitive operations must be performed. This is why a computer is called on to prepare an electric utility's 120,000 bills each month, to schedule a university's 38,000 students, to prepare a corporation's 17,000-employee payroll, and to process the Census Bureau's statistics on millions of citizens. Understandable. The key here is repetitiveness: volumes of data, millions of calculations; cyclic execution of the same basic operations on each piece of data.

3. **When only the computer is fast enough.** In the preceding types of computer applications, man and machine competed for the task at hand, and the computer got the job because it excelled at the task in question, on economic and perhaps other grounds. But there is no choice here; man cannot respond fast enough. Here is such an instance.

What could be an ICBM is detected by radar, apparently headed toward the United States. There is no time to spare. The nature of this object must be verified and a decision made concerning interception in a matter of seconds. Unfortunately, there is no time to fly out and have a look at the object for identification purposes. Those days are long gone. The only practical way now is to track the object by radar and on the basis of this information compute its location, speed, and probable destination, and then quickly see whether it can be accounted for on the basis of aircraft and spacecraft scheduled to be in the area. Based on this information, a choice must be made: forget it, wait, intercept. All of this must be accomplished in a few seconds; only a computer is capable of such speed. If an interceptor is to be launched, aiming it constitutes another challenging problem that only the computer can handle (Figure 13-10).

Conspicuous in this type of computer application is the speed

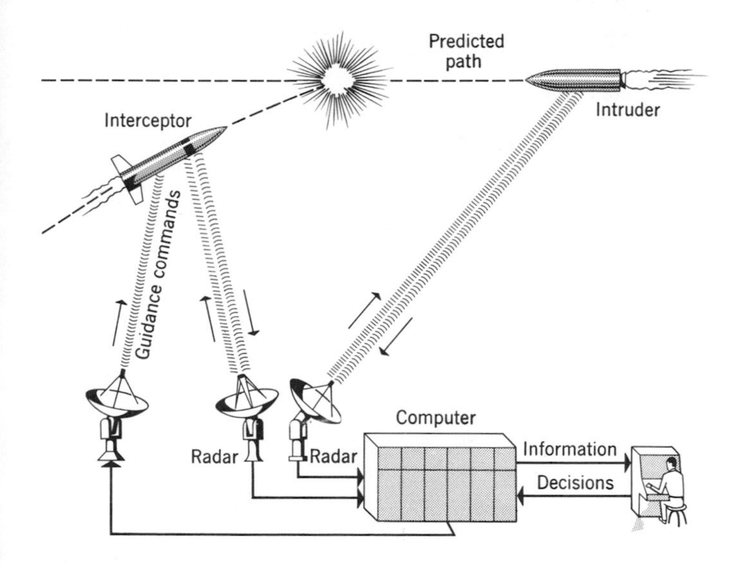

Figure 13-10 *To intercept and kill an intruding ICBM, the path of the intruder must be predicted and an intercept point calculated, which is likely to be hundreds of miles from its present position. Hence the interceptor has a computed path that it must follow to the point of kill. Unfortunately, there are forces that tend to steer it off course. Therefore it is necessary to sense deviations from the intended path, compute necessary corrections in course, and transmit these guidance instructions to the interceptor. This sequence of events–sense position, compute corrections, transmit–goes on continuously. This guidance system must react in microseconds in view of the missile's 29,000 km/hr speed. While all this is going on, the radar continues to sense the intruder's path and target. Therefore, the interceptor must be guided to a point of kill that, itself, is changing. And that's not all. An ICBM would no doubt be accompanied by clouds of decoys. The real things must be sorted out in a hurry. Such matters complicate the whole process considerably. One thing is for sure: buildings full of men could not possibly make these computations in time.*

with which things happen. Certain manufacturing processes like the rolling of steel slabs into flat sheets operate at very high speeds, making it impractical for humans to control them. In an increasing number of such instances, computers are called on to do the job.

4. **Too many things to keep track of.** Suppose there are 30 planes in the air around an airport. How can a man keep track of them? He can't! But a computer can (see Figure 13-11). For all practical purposes the computer keeps track of all planes continuously and simultaneously. Actually, however, it concentrates on one plane at a time, switching its attention from plane to plane and updating its knowledge of the whereabouts of each one at more

Flight A715

	$T-2$	$T-1$	T	$T+10$	$T+20$
x	9.12	9.03	8.94	8.49	8.04
y	27.04	27.03	27.02	26.97	26.92
z	2.14	2.14	2.14	2.14	2.14
	Positions 1 and 2 seconds ago		Position now	Predicted positions	

$T = 9{:}22{:}02$

Flight A715

	$T-2$	$T-1$	T	$T+10$	$T+20$	$T-2$
x	9.03	8.94	8.85	8.40	7.95	8.94
y	27.03	27.02	27.01	26.96	26.91	27.02
z	2.14	2.14	2.13	2.08	2.03	2.14

$T = 9{:}22{:}03$

Figure 13-11 *Once a second the computer receives information from radar on the position of each plane in its traffic control region. It stores this information in the form of* X-Y-Z *coordinates. What you see here is the computer's "file" on Flight A715 for two successive seconds. The computer is doing likewise for other planes under its control at the time, switching its attention rapidly from one plane to the next. The computer can do more than simply store positions. For example, it can predict where each plane will be in the near future, as shown in the predicted position columns. It does this by using the present and immediate-past positions of the plane to compute its speed and heading. Then on these bases it computes where the plane will be 10 seconds and 20 seconds later. In this way the computer can foresee potentially hazardous situations and alert traffic control personnel. Traffic controllers can request the computer to display this information graphically on their scopes, including the predicted positions, for specific planes or all planes in their area.*

than satisfactory frequency. This is the high-speed scan process that is characteristic of time-sharing computer systems.

Some manufacturing processes parallel the air traffic control situation in that they involve dozens of variables (temperatures, pressures, speeds, etc.) that are subject to frequent change and are interrelated. No human or any number of them can cope with all of these at once. Can you imagine what is involved in trying to keep track of 125,000 freight cars in a railroad system? Conspicuous in these situations is a large number of changing, interrelated events or conditions—too much for a human mind to grasp at one time. For the computer, this is no problem.

And so, in some instances, there is no choice; only a computer can handle the job. But in most applications it is a matter of effectiveness—alternative methods have been compared, and the computer was found to be superior. The designer of the utility company's billing system compared the total costs of preparing bills by desk calculators, accounting machines, and computer and found the latter to be the cheapest.

If you have become familiar with pages 225 to 236, you have a good overview of the manner in which engineers in particular and man in general are utilizing digital computers. With this overview in mind, you are better equipped to spot a potentially profitable computer application when you are staring at it.

Engineering Will Never Be the Same

Directly practicing engineers have felt the effects of the computer in two ways. For many it has meant learning new skills: numerical methods (a branch of mathematics rarely applied before computers) and computer programming. For most it has meant another type of knowledge to acquire—a general understanding of the nature of computers, their capabilities, limitations, and uses; the kind of information sampled in this chapter.

Indirectly, the computer has affected the practitioner in a variety of ways. For one, it has lead to expanded use of mathematics. Before computers, many sophisticated mathematical techniques served mainly as vehicles for mathematicians to demonstrate their prowess in classrooms and textbooks. Practical use of these techniques was severely restricted because of the man-hours required to solve the equations. But that obstacle has been removed. A whole range of powerful mathematical techniques can be applied; the computer will do the "dogwork" quickly and at a tolerable cost.

The computer is cutting routine, repetitive, tedious work to a minimum, a welcome change, indeed. Not that engineers no longer must do pencil-and-paper calculations or draw or search through files. No such luck. But it's certainly not like it was in precomputer days.

The labor-saving ability of the computer has another benefit. Before computers, practitioners were often forced to make gross, undesirable simplifications in many of their mathematical and simulation models. There is a very practical reason for this simplification: to yield equations that can be solved and simulation models that can be manipulated "by hand" in a reasonable period of time. This is a practical matter when you are under pressure to solve a problem as soon as possible, which is typical. An engineer who makes a business of designing structures has repeated need to predict the stresses in components of those structures. For this purpose there is a fancy mathematical model available, but a half day of hand computation is required every time it is used. He also has a simplified model that will require 5 minutes of hand computation. You know which model he is going to use, even though the simpler model is less accurate. Of course, he will compensate for the uncertainty in its stress predictions by using relatively large safety factors in his design. However, large safety factors mean increased material consumption (wider columns and beams in construction) and underutilization (a crane that could carry three tons with negligible likelihood of failure is limited to one-ton capacity). But these days, with the computer, he can use the more sophisticated model and benefit from its more accurate predictions and the correspondingly slimmer factors of safety.

The computer is helping to make better use of what is known

through its literature-searching and file-searching capabilities, although we have only scratched the surface in this respect.

Finally, the computer has greatly extended the engineer's "ability to accomplish." Many of man's achievements in space travel, nuclear power, air transportation, and communications would be impossible or long delayed without computers. This is worth noting, since laypersons are inclined to view the computer primarily as a displacer of workers; but after you have learned what the computer has made possible in engineering, medicine, government, architecture, literary research, and education, for instance, you are inclined to view the computer mainly as a doer of the heretofore impossible and impractical.

Incidentally, what engineers find the computer means for them, scientists have also experienced, and physicians, businessmen, lawyers, teachers, and others are beginning to experience. They, too, will find the machine gradually taking over the tedious and routine and vastly improving their "ability to do."

The Future

The contemplated effects of time-sharing on engineering offer a hint of the impact computers are going to have. Visualize an engineering organization with dozens of engineers distributed through many offices and buildings. They are served by a time-sharing computer with 40 input-output terminals, located so that every engineer has convenient access to the machine. All stations include a typewriter input to the computer, some have graphic equipment similar to that appearing in Figure 13-5, some have tape and card readers, and some have plotters and printers. From his terminal an engineer can initiate literature searches, call for information from files, solve equations, and do almost anything that is to be performed by computer. He can prepare a computer program, try it, "debug" it right then and, when he has a workable program, instruct the computer to store it. Anyone wanting to use this program can call for it from any terminal and have it available instantly.

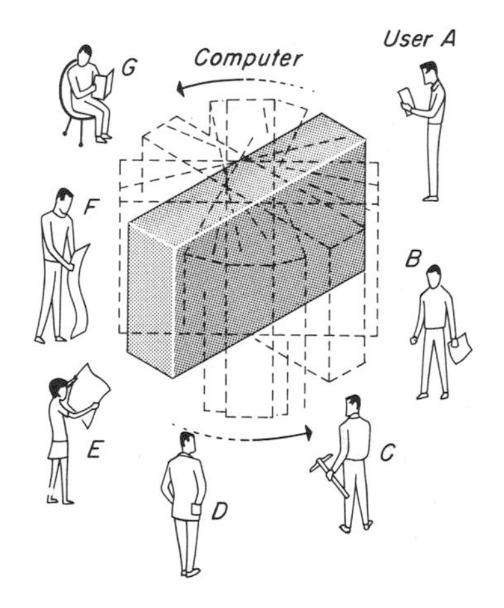

This system can have terminals thousands of miles away, in branch offices or other divisions of the company, tied into the one large computer by telephone lines. Such company-wide sharing of one computer gives everyone the use of a larger, more powerful machine, as opposed to buying a number of smaller machines and distributing them around the company. It also enables all engineers throughout the company to share technical information and computer programs. Such systems are around but far from commonplace.

Much of what engineers do with computers in the future is an outgrowth of time-sharing. It would hardly be feasible for me to sit all morning at a graphic terminal, sketching alternative designs, calling for different views of my drawings, and making computations if my activity required the complete attention of a million-dollar computer. But it doesn't; in 4 hours at the terminal I actually used 12 minutes of computer time, in the form of many spurts of fractions of a second. During the remainder of the 4 hours, the machine is available to the other 39 terminals.

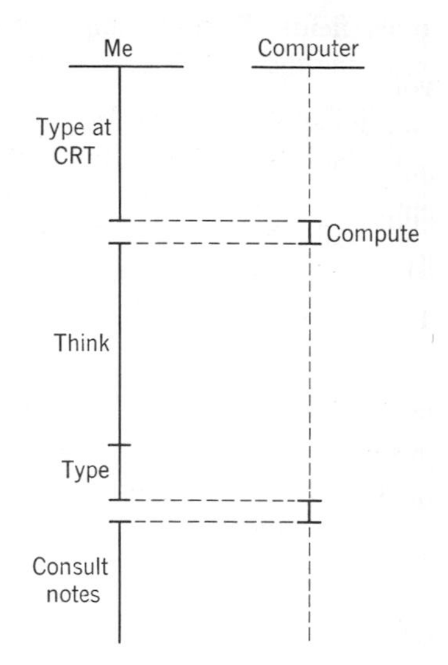

It is useful to view the engineer and the computer as a partnership in which man and machine complement one another, performing the functions for which each is best suited. Man excels at invention, at reasoning, at pattern recognition, and at learning from experience. He adapts quickly to a remarkable variety of tasks. He excels at the relatively short, one-shot task because his "setup" time (e.g., time to get pencil and paper in hand) is usually short compared to the time it would take to instruct a computer to do it.

In contrast, the computer performs repetitive, routine tasks reliably and precisely, without boredom or fatigue, in about a millionth of the time required by human beings. It has to be instructed only once, and thereafter it follows those instructions any number of times without deviation. It has a perfect memory for endless details and a memory uncluttered by useless information; when told to forget, it does so instantly and completely.

As improvements are made in computers and their programs, the machines will relieve engineers of more of the repetitive and routine tasks, allowing more time for activities consistent with their training. Thus the boundary between what humans do better and what computers do better is gradually shifting, to the engineer's benefit.

EXERCISES

1. A problem that concerns a number of people is the sizeable gap between the frontier of knowledge in a particular field and what the practitioner typically knows and uses. Mankind is not particularly effective in putting what it learns into widespread use, and this is true of education, medicine, manufacturing, and

other fields. Digital computers offer considerable potential in reducing that gap. Can you visualize how?

2. Patterning your response after Figure 13-9, summarize what you visualize a large-scale, time-sharing computer system with ample auxiliary storage could do for a city government (or a university, or the police department of a large city, or . . .). The emphasis here is on what the computer *can* do; economic feasibility and social desirability are other and more complicated matters.

3. What is your reaction to the digital computer? Do you see it as boon or bane? Cite some specifics in your argument. Your response to this question can be based partly on Chapter 13 and perhaps mainly on your general knowledge of computer applications and impact.

4. Certainly a digital computer contributes much to the engineer's modeling capabilities. Scan Chapters 9 and 10 and cite where and how the computer enhances this capability.

5. A digital computer has been a real assist to the engineer in his optimization efforts. Scan Chapter 11 and identify where a digital computer would be useful.

6. In general, what do computers have to offer to feedback systems? Don't restrict your thinking to engineering systems; consider health-care, business, educational and other types of control systems, too.

SOME GENERALIZATIONS

Chapter 14

Before generalizing, here is one more brief example of the types of things engineers must know. In this case, it is what is required to design a transformer.

A transformer generally consists of two current-carrying coils electrically insulated from each other and wound on the same iron core (Figure 14-1). An alternating current in the primary coil produces an alternating field of magnetic flux that passes through the metal core to the secondary coil in which an alternating current is induced.

In a basic physics course an engineering student usually learns something about transformers, under the subject of electro-

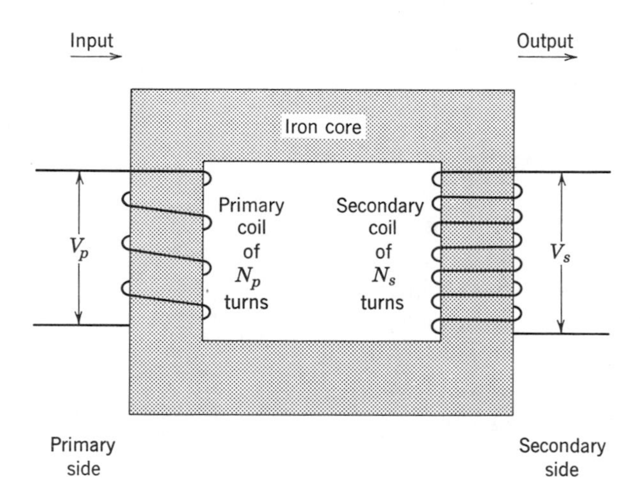

Figure 14-1 *A simple step-up transformer. It transforms the primary voltage (V_p) to a higher secondary voltage (V_s).*

magnetism. However, the coverage is ordinarily limited to the principles of ideal transformers. Probably the only quantitative relationship he learns is the one between the primary and secondary voltages, described by this simple model. But that's about all. If he is assigned problems to work he is given numerical values for three of these variables and asked to compute the fourth (e.g., given values for N_P, N_S, and V_P, find V_S). But the engineer must know considerably more than this to *design* transformers.

(*a*)

$$\frac{V_P}{V_S} = \frac{N_P}{N_S}$$

where
V_P and V_S are the primary and secondary voltages.

N_P and N_S are the number of windings in the primary and secondary coils.

For one thing, he must extend his knowledge beyond the ideal case. Therefore, in a subsequent applied science course, probably in his sophomore year, he will learn more about the types of losses that were assumed away in the study of the ideal transformer. There are realities such as resistance of the wire in the coils, flux leakage, eddy currents, and hysteresis, which collectively cause the output of a real transformer to be less than the ideal. Again, however, if he is asked to solve transformer problems in such a course, he will most likely be given numerical values for all but one of the variables, including losses, and be asked to solve for the one unknown. But certainly it is not that simple in practice.

On the job he will usually be told that a transformer is needed, with an output specified in terms of the current (I_S) and voltage (V_S). He may also be told the primary voltage (V_P) and the alternating current frequency (*f*), but not much else. He takes it from there. But many of the variables like N_P, N_S, and I_P are as yet unknown; so are the size and shape of the core, the material from which it is to be made, and a number of other unspecified solution variables. *This* is a lot different than the relatively simple "plug in all the values but one and crank out the answer" routine that sufficed in introductory courses. Here the engineer has many interdependent unknowns; how can he assign values in such a situation?

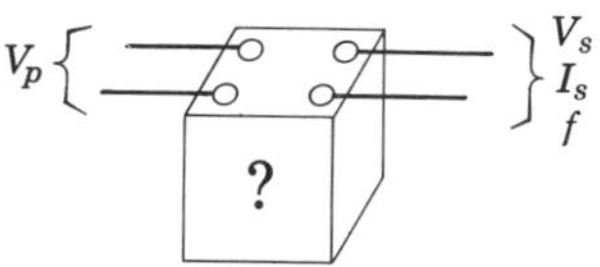

Given this incomplete information, the engineer must determine all other characteristics

The details of his procedure are not important here but, in general, it goes like this. The designer has this empirical equation, available in an electrical engineering handbook, that enables him to estimate the number of windings needed in the secondary coil, *if* he is willing to make some assumptions. So, for a starter, he must assume trial values for *A* (the cross-sectional area of the transformer's iron core) and *B* (the density of the magnetic flux flowing through the iron core). There are rules of thumb in practice to guide his selection of trial values for *A* and *B*. Although these are only "ball-park" values, they do enable the engineer to get a foothold in the cyclical process of successive approximations.

(*b*)

$$N_S = \frac{(V_S)\ 10^8}{4.44\ BAf}$$

Now, using the values given for V_S and f and the values assumed for A and B, he can use equation (b) to compute a first approximation of N_S. Now he has values for V_S, V_P, and N_S, and using these in equation (a), he can calculate N_P.

After making tentative selections of materials and the coil-core configuration, he can estimate the resulting losses. But note that many of these variables are interdependent; B, for which he originally assumed a value, depends on many of the subsequently established characteristics. So about now he starts the process over again, but this time with a better basis for setting values of A and B. Thus he goes through a series of successive approximations until all features appear satisfactorily specified.

Even though you may not follow all of this, it should be apparent that there is much more involved than the simple plug-in process associated with the use of equation (a) in introductory physics. There are few equations and many interrelated unknowns, so that design of a transformer turns out to be a cyclic, cut-and-try process, in which the engineer relies heavily on empirical knowledge in the form of formulas, rules of thumb, tables, and graphs. Some of this empirical information is acquired from upper-class engineering courses, but most of it comes from on-the-job experience and handbooks.

Some Generalizations

Basic Physical Science

A very important part of an engineering education concerns the physical sciences, primarily physics and chemistry, as evidenced by the number of courses in these subjects in the typical engineering curriculum. In order to develop complex devices, structures, and processes, you must acquire a *fundamental* understanding of the laws of motion, the structure of matter, electromagnetism, the behavior of fluids, the conversion of energy, and many other phenomena of the physical world.

But knowledge of basic physical science is hardly enough. It takes a lot more than familiarity with ideal transformers to design real ones successfully. This is true for all engineering creations. People don't want their bridges designed by engineers who know only basic physics and chemistry any more than you want to be treated by a physician whose only qualification is knowledge of basic physiology. There is a big step between the basic principles of physical science and technical solutions to real problems. Your formal education must equip you to bridge this gap. Therefore, after the rigors of basic physical science, you must study *applied* physical science, engineering systems, and an accumulation of empirical know-how.

Applied Physical Science

One thing you must do before you can beneficially apply scientific principles to the world's practical problems is study *applied science*—the "how" and "where" of applying basic principles. An example of applied physical science is circuit analysis, which concerns the application of knowledge of fundamental electrical phenomena (charges, electromagnetic waves, electron flow, etc.) to the understanding of basic electrical circuits. It is in such a course that the engineering student extends his knowledge of transformers from the ideal to the actual case. Other applied physical sciences are taught under course titles such as thermodynamics, mechanics of solids, fluid mechanics, and properties of materials. These are concentrated in the middle portion of the typical engineering curriculum.

Engineering Systems

Engineers are creators of systems ranging in size from microsystems like the electrical circuits seen only through a microscope to the supersystems exemplified by our telephone network. The transistor, toaster, automobile, computer, airplane, dam, oil refinery, highway network, and all other engineering creations *are* systems. To design such systems, engineers must understand them thoroughly and know how they are synthesized. Thus, in courses in their junior and especially their senior year, they have learned to design communication, transportation, energy conversion, control, manufacturing, and other types of systems.

However, no student studies *all* of these. It is customary to specialize to some degree, primarily because large and substantially different bodies of knowledge are required by basically different types of problems. It is virtually impossible for an engineer to be competent in designing bridges *and* television equipment *and* jet engines *and* metal refineries *and* textile machines. As a consequence, some specialization is inevitable. Therefore, during the latter part of your undergraduate program, you will major in some branch of engineering. The choices are many; some major ones of longstanding importance are:

- **_Aerospace_ engineering. Primarily the design of systems for travel above the earth's surface. Examples: aircraft, spacecraft, air cushion vehicles, guidance and other systems associated with flight.**
- **_Chemical_ engineering. Primarily the design of processes for the chemical transformation of materials on a large scale. Examples: facilities for the production of gasoline, paint, explosives, rubber, cement.**
- **_Civil_ engineering. Primarily the design of major structures _and_ the means of constructing them. Examples: highways, bridges, dams, canals, water supply and sewage disposal systems, airports, and waterports.**
- **_Electrical_ engineering. Primarily the design of means by which electrical energy is created, transferred, and used. Examples: electrical generators, transmission networks, communication systems.**
- **_Industrial_ engineering. Primarily the design of operating systems for producing goods and services. Examples: automobile plants, printeries, shipyards, hospitals (not the building itself).**
- **_Mechanical_ engineering. Primarily the design of systems by which energy is converted to useful mechanical forms. Examples: combustion engines _and_ the mechanisms required to convert the output of these machines to the desired form, including pumps, compressors, and transmission systems.**

It is primarily in the third- and fourth-year design courses that education in the branches of engineering differs. The student of electrical engineering studies the behavior and design of electrical machines, communication systems, power distribution networks, and the like, while the student of civil engineering studies structures, water supply systems, city planning, and related subjects. One type of system that they all study, regardless of specialty, is the feedback control system, because most engineering creations include such systems.

Although specialization along traditional lines is still common in engineering *education,* most problems encountered in practice require knowledge from two or more of the traditional engineering specialties, as demonstrated by the case studies in Chapter 1. Design of a commercial chemical process such as a plant for converting oil-sands into crude oil requires knowledge that is traditionally a part of the education of a chemical engineer, as well as knowledge acquired by civil, electrical, industrial, mechanical and metallurgical engineers. As a result, an engineer must often work closely with engineers educated in specialties other than his own, and must himself employ some of the knowledge from other branches of engineering. Thus, the engineer typically finds that, on the job, his knowledge must extend across the traditional specialty boundaries. Mainly for this reason you probably will be required to take some courses in engineering specialties other than your own.

Codified Empirical Knowledge

This is an accumulation of ideas, standard practices, formulas, rules of thumb, and the like, some of which is published and some of which is retained in heads, files, and notebooks. The nature and role of this practical knowledge is apparent in the transformer case. Such information is not based on scientific principles. It is cumulative know-how that has evolved over a long period, during which it has proven sound and generally useful. Engineers rely heavily on this "communicated experience," and so this type of knowledge is woven into upper-class engineering studies.

Other Knowledge

There are some important nontechnical aspects of your intellectual development; certainly knowledge of physical science and engineering will not suffice. The practice of engineering requires some familiarity with economics, political science, psychology, and sociology. This breadth of knowledge is important for a number of reasons; for instance:

- **You must know the "economic facts of life." To be of value to employers and to benefit society you must be aware of the importance and intricacies of costs, price-demand relationships, return on the investment, depreciation, and other economic realities. You will be heavily involved in economic decisions.**
- **You will be working with persons in many fields of**

endeavor: economists, accountants, politicians, sociologists, psychologists, lawyers, and union leaders. You should be aware of the contributions that these people can make, be able to talk with them intelligently, work with them, and understand their problems.

- **Educational breadth equips and motivates you to show greater concern for the society you affect. What better reason is there for insisting that your education not be exclusively technical?**

For reasons such as these, a minimum of one-fifth of an undergraduate engineering education is reserved for study of the humanities and the social sciences.

In summary, an engineer's creations are based on an accumulation of knowledge varying from pure science to pure empiricism, acquired in science, applied science, and engineering-systems courses, as well as through experience. In general, the typical solution owes less to science and more to other types of knowledge than many authors would lead you to believe. It is worth making an issue of this, partly to refute the common impression that engineering is (or is primarily) the application of science. Nonsense! Of course, engineers use science in the solution of problems, but much of what they rely on is empirical (recall the transformer case) or is under the general heading of engineering systems. Furthermore, those who consider engineering to be applied science are ignoring the fact that engineering is still very much an art. It is partly to squelch the view that engineering *is* the application of science that the following is offered.

The Origins of Engineering

Man has always devoted much effort to the development of devices and structures that make natural resources more useful. He devised the plow to render the soil more productive, the saw to transform the wood of the tree to useful forms, the windmill to convert the forces of the winds to useful work, and the steam engine to transform latent energy in fuels to useful work. These and myriad other implements, machines, and structures are results of an unceasing search. In early times, as occupational specialties were evolving, there arose along with priests, physicians, and teachers a specialty devoted to creating these devices and structures. These early engineers were responsible for creating weapons, fortresses, roads, bridges, ships, and other contrivances. Their activity is readily traceable to the ancient empires and evidence of their remarkable creations still remains.

These men were the predecessors of modern engineers. *The most significant difference between these classical engineers and their modern counterparts is the knowledge on which their creations are based.* The early engineers designed bridges, cathedrals, machines, and other major works using experience-based practical know-how, common sense, experimentation, and inventiveness. The practical know-how was an accumulation of experience passed on mainly by the apprenticeship system, to which each man contributed from his own experience. In contrast to today's engineers, their knowledge of science was meager. This is no reflection on them; at the time, man's understanding of the physical world was crude, indeed.

Engineering remained essentially in this form for many centuries. During the Renaissance the level of sophistication increased but, even in the eighteenth century, the creators of machines and structures relied mainly on empirical knowledge and invention and very little on science. Evolution of the steam engine illustrates the state of engineering during that period. The steam engine patented in 1769 by James Watt was one in a succession of progressively better machines, which began more than a century before. Watt made a significant improvement that vastly improved the efficiency of the steam engine and finally led to its widespread use. The evolution of his machine is marked by a series of cumulative inventions contributed by men who knew little or nothing about the scientific principles underlying their accomplishments. They knew nothing about molecular activity, the quantitative relationships between temperature and pressure, heat transfer theory, and the like.

The classical engineers were handicapped in what they could accomplish as long as they had little basic understanding of energy, motion, materials, and the like, a situation that existed until relatively recent times. All this has changed. Within the last two centuries, scientific knowledge has blossomed into an immense accumulation of information. Man's understanding of the structure of matter, the elements and their relationships, the laws of motion, energy transfer processes, and many other aspects of the physical world has improved many times over. Most of what is now taught in high-school physics was unknown when Watt developed his steam engine. Yet the contents of such a course are only a fraction of what is known today.

Actually, it was not until the late nineteenth century that significant amounts of scientific knowledge were *applied* in the solution of engineering problems. The trend since then has been to rely on it more and more. With this major change—the increased use of scientific principles on the solution of problems—classical engineering evolved into the engineering of today.

So contemporary engineering is primarily the outgrowth of two historical developments that, until relatively recently, were essentially unrelated. One of these was the evolution ages ago of a specialist who has since served as society's expert in the creation of complex devices, structures, machines, and other contrivances. The other development is more recent: the accelerating growth of scientific knowledge. Although the marriage is a fairly recent one, it has already brought about a significant change in engineering. In contrast to engineering of the past, modern engineering involves more science, although art is still very much in evidence.

There is a close parallel between the evolution of engineering and that of medicine. Specialists in the curing of bodily ills evolved in very early times. These predecessors of modern physicians practiced through many centuries what was essentially an art; there was no significant body of scientific knowledge on which to rely. In relatively recent times bacteriology, physiology, and other biological sciences developed into respectable bodies of knowledge, and the physicians began to apply them in the treatment of human health problems.

So both physicians and engineers are problem-solving specialists with roots deep in history, who eventually and logically assumed the responsibility for applying a certain body of scientific knowledge. *They always were and still are problem-oriented*. Their prime motive is to solve a problem at hand. If perchance they are faced with a problem and there happens to be no scientific knowledge that tells them what to do in that instance, they still attempt to solve that problem. (The surgeon does not walk away from a patient on the operating table when he encounters a situation for which science does not tell him what to do!) They have a job to do and, if necessary, they arrive at a solution to the problem at hand through experimentation, common sense, ingenuity, or perhaps other means, if current scientific knowledge does not cover the situation. Thus, engineers do not exist *solely* for the application of science; instead, their goal is to solve problems and, in so doing, they utilize scientific knowledge when it is available.

They do what they must; use science when applicable, intuition when useful, and trial and error when necessary.
Charles L. Best

Once you are aware of the origins of engineering (i.e., the

practical arts of early civilizations, millennia before science had anything significant to offer to engineers), you are able to dispel one of the common misconceptions concerning engineering—the "offspring of science" myth. But there are other misconceptions that arise out of the close relationship between science and engineering; the matter is worth exploring further.

Science and Engineering–The Distinction

True, science and engineering are closely related and, in some respects, are interdependent. On the other hand, there are some fundamental differences between them that elude many people partly, no doubt, because of this very relationship. Science and engineering differ in four major respects: *the basic processes characteristic of each; predominant day-to-day concerns; primary end-product;* and *knowledge employed.*

Science is a body of knowledge, specifically, man's accumulated understanding of nature, *and* the means by which that understanding is improved. *Scientists* direct their efforts primarily to extending this knowledge. In their search for better understanding, scientists engage in a process called *research.* In so doing they devote much of their time to hypothesizing and refining models (explanations, classification schemes, mathematical representations, etc.) of natural phenomena and to evaluating those models on the basis of observations from experiments and the world around them (see Figure 10-3). There is certainly more to the work of scientists, but the intent here is mainly to highlight their heavy involvement with models: model development, model refinement, and model verification. Now contrast this with the engineer's design process in which modeling is strictly a means to quite different ends.

These ends, in contrast to knowledge, are tangible, generally in the form of physical systems created to satisfy some human need or want . Let there be no mistake—the weather satellite, the radio telescope, the heart-lung machine, the nuclear power station, the electronic computer, and the artificial kidney evolved from engineering endeavors. Engineers develop these contrivances through a creative process referred to as *design* (in contrast to scientists' central activity—research). Some of our prime concerns as designers are economic feasibility, safety, public acceptance, and manufacturability of solutions. These are hardly major concerns of scientists. Instead, scientists are more concerned about validity of their theories, reproducibility of their experi-

Figure 14-2 *This is by far the largest radio telescope. It is a wire mesh reflector supported in a natural hollow in the mountains of Puerto Rico. Radar signals originate from the movable transmitter and are reflected toward outer space. These signals bounce off planets and stars and return as echoes that are focused by the reflector on the suspended receiver. These echoes are then analyzed by scientists to yield new knowledge about the universe. Here is a prime example of an engineering creation that the press typically calls a scientific achievement. To be precise, it is a scientific instrument used in radio astronomy, the design and construction of which are engineering achievements. Engineers determined the site for this telescope; they designed it; in fact, an engineer conceived the basic idea. I mention these things not because scientists and engineers quibble about credit due, but because the types of work involved in design and use of this instrument are quite different, and this is important to persons planning their careers.*

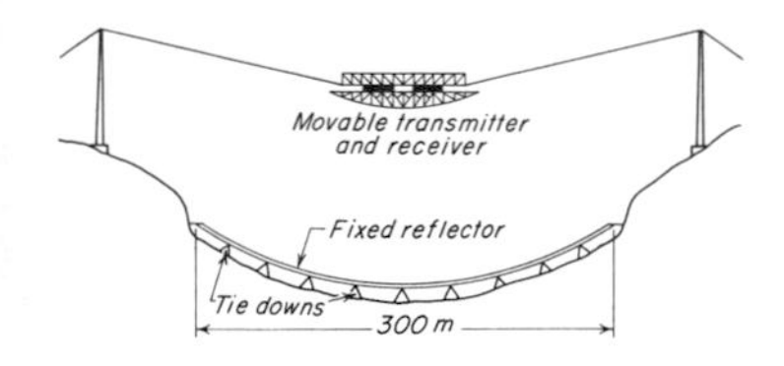

ments, and adequacy of their methods of observing natural phenomena.

But it's not only objective, methodology, and day-to-day concerns that distinguish these two disciplines; even their knowledge needs differ. This can be illustrated by contrasting what scientists and engineers must know about control systems. Scientists, particularly in the biological and social sciences, should have some knowledge of control systems, but certainly not because they design them. Their primary interest is in *explaining* existing control mechanisms in nature (e.g., the means for controlling blood supply in the body). In contrast, the prime objective of engineers is to *create* systems that fulfill control needs in man-made systems (e.g., *given* that steam pressure is to be maintained at a specified pressure, to design a system that does so).

Similarly, a comparison of what a student learns about how transformers operate in a physics course and what an engineer must know to design them can be instructive. Note that in a physics course a student learns how to determine what the result will be, given various voltage, current, coil, and other conditions. In contrast, the engineer must determine characteristics of the coil, input, and the like, *given the desired result (output)*. Thus, in engineering the emphasis is on *creation,* whereas in science the emphasis is on *explanation* (some like to express this crucial difference in terms of a synthesis and analysis).

Scientists explore what is and engineers create what has never been.
Theodore von Kármán

Faraday's formulation of the principles of electromagnetic induction was a contribution to science. The use of this knowledge in the design of electric generators is engineering. When man came to understand nuclear fission in the late 1930s, this was an important scientific discovery. Applying this knowledge in the design of power-generating nuclear reactors is engineering. This is not to say that engineers do not do research in the course of finding solutions to their problems. Engineers who developed practical means of converting brackish water to usable form performed some research to gain additional knowledge about the fundamental processes involved. However, their goal was development of an economical conversion process; they engaged in research only in order to solve that problem. When a space vehicle reenters the earth's atmosphere at very high speeds, the heat generated will melt any known metal. It was necessary therefore for engineers designing reentry vehicles to engage in research to find a material capable of withstanding this intense heat. The knowledge that resulted is a by-product of their efforts to develop a successful reentry vehicle.

So engineers do research, and certainly scientists design instruments and solve problems. The key to the distinction is what constitutes a prime objective and what constitutes a means to an end.

Turning from Widely Held Misconceptions to a Little Recognized Truth

While there are some myths surrounding engineering, there are also the inevitable little-known truths. The next chapter is addressed to a very important one: the role of the engineer as an agent of social change, a fact that seems to elude many engineers as well as the public. Yet, through their creations, they precipitate more social change than those who work at it, although they do so indirectly, unintentionally, and sometimes unwittingly. (The efforts of an army of suffragettes never did have the impact of one relatively small engineered device—the self-starter for automobiles, which made it practical for women to drive.) This role as a social revolutionary—precipitator of social transformation—is worth examining in detail.

EXERCISES

1. Take a nuclear power station, a sewage treatment plant, a submarine telephone cable, and a petroleum refinery and identify the types of engineering specialties that surely are involved in their design. You can base this analysis on the text, a little extra reading, and common sense.

2. Talk to an engineering senior and get a firsthand explanation of the nature of his studies, making sure to touch on the major categories outlined in this chapter. Submit a short report summarizing your conversation.

3. Assuming that you are in engineering school, each course that you are now studying is intended to contribute to your development with respect to certain of the qualities described in this chapter and in Chapter 2. Analyze the content of each of these courses; isolate the types of knowledge, skills, and attitudes that the course is apparently intended to develop in you.

4. Prepare a list of devices, structures, and processes created by each of the major branches of engineering described.

5. Prepare a list of 15 engineering creations, the design of

which probably required the talents of engineers from two or more major branches of engineering. Identify the branches that you believe were involved in the development of each creation that you name.

6. A mental exercise effective in clarifying the distinction between scientist and engineer is some reflection on the training, primary missions, predominant activities, and characteristic problems of a research chemist and a chemical engineer. A little extra reading and conversations with such persons will be helpful. Supply evidence of your reflections.

7. In this chapter and elsewhere in the text you have read of some prevalent misconceptions concerning engineering and engineers. Summarize those *and,* more important, cite some misconceptions based on your own experience and observations.

8. Figure 14-2 describes a prime example of an engineering creation commonly described as a scientific achievement. Can you cite other examples?

9. What is your reaction to the following definition? "Engineering is the application of science."

Engineering and Society—The Interaction

PART 4

Part 3 is an introduction to some of the knowledge you will need as a *practitioner*. Part 4 is an introduction to some of the knowledge that you will need as a *professional*. It is concerned mainly with the nature and significance of your interaction with society. Engineers do not operate in a vacuum, far from it. Their jobs, attitudes, and contributions are strongly affected by the economic, organizational, social, political, and legal environments in which they operate. Engineers, in turn, primarily through their creations, significantly affect employers, consumers, the public at large, and a variety of economic, social and political institutions. This multifaceted interaction is the subject of the next four chapters.

THE IMPACT OF ENGINEERING

Chapter 15

Tools, machines, and structures have had far-reaching effects on mankind; the evolution of these contrivances has been closely intertwined with major political, social, military, and economic events throughout history. Visualize, for example: the commercial, cultural, and political interactions between nations that resulted when vessels capable of traveling the high seas were developed; the increase in agricultural productivity when the iron plow replaced its wooden predecessor; and the impact of the printing press on the preservation and dissemination of knowledge.

The creations of modern engineers are no different in this respect. The automobile obviously has significant social, economic, and environmental effects. Agricultural mechanization has greatly hastened our transition to a predominantly urban society. Mammoth power plants convert fuels into energy that, in turn, underlies our industrial and agricultural might; this, in turn, has enabled us to become a society with an unprecedented standard of living. And is it necessary to mention "the bomb" and its impact on the affairs of men? Or automation? Or the computer?

This chapter explores the impact of engineering at two levels. The first, the obvious, and the simplest to explain involves the effects of an *individual* creation (a giant machine for irrigating the desert, the snowmobile, or an emission control device for automobiles). The second level of impact involves the *collective* effects of many creations, generally a certain class of them (the aggregative effects of modern communication media, the total effect of all machines on the atmosphere, or the impact of all appliances and other machines on family life). This type of impact is more elusive; it is generally difficult to perceive, let alone to influence. Consequently, the impact of the collective works of engineers will receive proportionately more attention.

Impact of a Specific Technical Venture

Consider a familiar, relatively straightforward case: the construction of a new freeway in a suburban area. You *know* that it visibly affects the landscape, that it means added convenience (and safety, by the way) for millions of drivers, that it is noisy and smelly, that it displaces many families and disrupts long-established neighborhoods, that it influences land use for miles around for a long time to come, that it affects established drainage patterns, and that it drastically alters traffic flows on connecting roads and nearby streets. Surely, if you were locating and designing this freeway, you could not escape an awareness of these and other effects of your creation.

However, for every case like this in which the far-ranging effects are relatively apparent, there are many technical ventures for which the full range of effects is obscure. This is true in the case of nuclear energy, the plan to redistribute Arctic waters over North America (page 52), satellite transmission of television over a multination territory, aerosol sprays, or an offshore tanker terminal.

The snowmobile might strike you as a relatively benign development and, as far as the United States is concerned, this may prove to be so, although some naturalists would surely disagree. However, in areas further north, the effects go far and deep. In northern Canada and especially in northern Scandinavia, where reindeer herding is a primary industry, the arrival of snowmobiles has meant, in addition to the environmental effects:

- **A significant increase in social interaction.**
- **Better medical aid and other services.**
- **A new industry associated with the sale and servicing of these machines.**
- **Doom for beasts of burden; sled dogs and reindeer are becoming obsolete.**
- **Larger harvests by fishermen, hunters, and trappers. The quantum jump in mobility provided by the snowmobile resulted in a virtual elimination of bears in one season in Finland.**

- **A major social problem that closely parallels what extensive mechanization has meant to small farmers in this country. The deer-herding industry is shifting from labor-intensive to capital-intensive, spelling economic doom for a majority of the small, marginal herders.**
- **Increased reliance on an external source of energy: oil.**

Those Pesky Indirect Effects

Note that many of the consequences of engineers' creations are indirect, such as the socio-economic changes precipitated by the snowmobile. Such consequences are often referred to as side effects, or second-order effects. Jacob Bronowski offers these tongue-in-cheek illustrations of side effects.

> We never know with certainty what the social consequences of any discovery will be. Who would have thought that the unfortunate character who invented photographic film would have been responsible for the California film industry? And thus, indirectly, for contracts that would prevent film stars from having affairs that might give rise to gossip and scandal? That consequently stars would lead their love life in public, by repeated divorce and marriage? That therefore the beautiful pin-ups of films would, in time, become the models of the divorce business? And the climax, that one-third of all marriages contracted this year in California are going to end in divorce—all because somebody invented the process of printing pictures on a celluloid strip?
>
> On the same lines (which I leave you to trace), who would have supposed that Henry Ford's devising of the sequential method of assembling a motorcar would finally result in upsetting a whole moral code of the American middle classes. For it is evident now that the car provided young people with more privacy than the home, and that as a result it became usual to begin sexual experience on the backseat of a motorcar.
>
> Although my examples may seem extravagant, they are not. The fact is that, in a strange way, the side effects of technical innovation are more influential than the direct effects, and they spread out in a civilization to transform its behavior, its outlook, and its moral ethics.*

* "What We Can't Know," *Saturday Review,* July 5, 1969, by permission.

Ordinarily, second-order effects are unintended; often they are unanticipated. On occasion they amount to a chain reaction of interrelated effects that would be virtually impossible to predict. Frequently the victims are "innocent third parties" who are neither instigators nor beneficiaries of the change. Television may have some adverse social effects, but the people who are thus affected are the same ones who use and enjoy the medium. But many of the persons who are displaced by new freeway construction are not beneficiaries of the new road; the same applies to factory workers who are displaced by new production equipment and office employees whose jobs are taken over by computers; and, too, many people who live near major airports and suffer the familiar adverse effects have never been in a plane. True, not all side effects are negative; there are second-order benefits as well as costs. However, indirect benefits of a technical venture, anticipated or not, are hardly cause for concern.

Second-Order Capabilities

Observe how engineered devices, machines, and structures underlie many of man's other important capabilities. Agriculture is an excellent case in point. It contributes what it does and employs relatively few people in doing so, mainly because of the variety and potency of the machines employed in most phases of the business (including in the large-scale manufacture of fertilizers and insecticides). The entertainment industry would scarcely have the impact it now has if it weren't for the widespread availability of motion picture machines and television facilities. Or consider medicine; without instruments and machines for diagnosis, surgery, and therapy, without artificial limbs and pacemakers, without mass production of drugs, how potent would medicine be? It is because of this "amplifying effect" of engineering creations in other areas of human endeavor that the term technology is so useful.

On the following pages it will be helpful to have a single term to refer to the totality of mankind's capabilities that derive directly or indirectly from engineering. That term is *technology*.* It embraces, for instance, not only television facilities, but all of the capabilities of the entertainment-communication complex that derive from that equipment. This same is true for all of the

* It is important that you know there are numerous interpretations of *technology*. To get some feel for the confusion surrounding this term, refer to the appendix to this chapter.

implements, machines, mass-produced chemicals, and the like that engineering provides to agriculture and medicine, *and* everything these make possible in those fields, and for what engineering contributes to manufacturing, commercial fishing, food processing, dentistry, warfare. . . . And, if you doubt that these capabilities are highly engineering-dependent, picture what remains when you take away the engineer's contrivances that underlie any one of those capabilities, such as medicine.

Observe, then, that the impact of engineering may be further reaching than you suspected in view of the powers engineering gives to others which, in turn, have their own direct and indirect effects.

Enough, for the moment, on the impact of engineers' creations on human affairs and human welfare. This brief sampling of a story that fills many volumes is intended to make you "impact conscious," generally sensitive to the manifold ways in which engineers' creations profoundly affect our physical comfort and safety; our power to control; our capacity to injure; our mobility; the ease with which we communicate; the education we need; our life span; the hours, content, physical demands, and stability of our jobs; our environment; our values; and our behavior. The next step is to explore the consequences of all this for engineers.

The Consequence for Engineers: Social Responsibility

Engineering creations directly affect people, usually many people and in many ways. *People* ride transit systems, *people* operate machines in offices and factories, and *people* service automobiles. The obligation to serve well the people directly affected by their creations is an important professional responsibility for engineers. But it is not the only one; society is not going to let them escape *some* responsibility for adverse side effects like the social disruption of communities by urban freeways. Thus, you are expected to recognize the *total* physical, economic, psychological, and social impact of your solutions, including the indirect effects, the unintended consequences, the unquantifiable results, and the implications for nonusers.

If you are to fulfill these responsibilities, you must come to "know" the society you affect and to care about its well-being. This "knowing" and "caring" requires some explanation.

In all of their projects engineers should learn what people need, prefer, and will tolerate, and this awareness should be reflected in their solutions. If the engineer designing a new transit system for a community is to satisfy the maximum number of people, he should know their needs and wants. Where do people want to travel? When? How much importance do they attach to privacy? To pickup frequency? To comfort? To other criteria? Are enough people willing to give up the privacy and independence afforded by automobiles to make a new transit system feasible? How do they feel about overhead train-carrying structures passing through suburban areas? What noise level do they consider reasonable? What do they consider esthetically pleasing? This is all part of getting to know the people directly and indirectly affected by the system. While designing it, he should use this information to predict how people will be affected by and react to alternatives he is considering (e.g., overhead versus underground, individual cars versus trains) in an attempt to maximize satisfaction and minimize adverse effects and reactions.

Predictions of the economic, social, psychological, and political impact of technical alternatives require special insight concerning man's individual and social behavior. For this to be, you had better learn something about these matters in college. And you will. But knowing society is not sufficient; you must also care. It is one thing for the designers of a large dam to not be oblivious of the psychological, social, and economic effects it will have on the farmers and villagers who will be dislocated. It's another thing to be aware *and* to be actively concerned about these effects by giving them due consideration in their benefit-cost deliberations, by minimizing the hardship created, and by facilitating the economic, social, and personal adjustments of the people affected. *That's caring!*

Engineers have been criticized for showing insufficient concern for the full implications of their creations. *Some* of this criticism is misdirected. Public officials and business executives make many of the decisions for which "the engineers" are unjustly blamed. However, while engineers are not to blame, we are not blameless; enough of this criticism is deserved to warrant some serious introspection.

True, engineers rarely have the ultimate say about what businesses and governments do with their engineering capabilities, but this hardly justifies not caring and not taking a tough stand on behalf of mankind. After all, we *are* as much creators of social change as of physical change; we *should* know and care about the victims as well as the beneficiaries of our creations. It *is* appropriate that our designs closely match human needs and wants, that we intelligently predict the full impact of our solutions, and that we minimize the adverse social effects and maximize the social benefits of our creations.

Few will disagree that you should learn about and care about the effects of your creations. However, what constitutes a proper professional response to the collective effects of engineers' creations—those described starting below—is not clear-cut. Certainly you can't be held accountable; on the other hand, it would hardly be appropriate for engineers themselves to be ignorant of the consequences of their collective works.

Collective Effects

Some of technology's implications are direct and obvious. You have no difficulty noticing; you *experience* many of them. Then there are other implications, at least as important, that may well escape you, and often these are the collective effects introduced earlier.

For example, the effects of machines sometimes converge on a particular segment of the population, as in the case of the blacks, to produce an especially unfortunate situation. Look how machines—automobiles, trucks, farm equipment, and manufacturing machines—have combined to create an employment crisis for blacks. Agricultural machines, such as those that now pick 97 percent of the cotton crop, have displaced thousands of black laborers on southern farms. In fact, they have put many operators of small farms out of business, and many of these are black. In the meantime, some equally consequential developments are taking place in the urban centers. For a number of reasons, some of which are technological, factories are migrating from the city to outlying areas. This is feasible because automobiles have made employees so much more mobile and because trucks have cut the ties between factory and railroad. The result is that employment opportunities are shifting from city to suburb, during a period in which there is a major influx of job-hunting blacks into center city, many of whom lost their jobs to agricultural machines. They feel the pinch of machines again when they find that low-skill jobs are becoming scarce as a result of continuing mechanization in factories, offices, and construction.

Certainly you have observed the collective effect of machines on the family, the housewife in particular. A small army of electro-mechanical slaves plus the automobile make the working mother much more feasible which, in turn, has acknowledged social and economic consequences.

Another, more subtle, illustration is provided by the mass media. The printing press, radio, motion picture, and television have made their marks individually, but they also have their collective effects. One timely instance is their role in precipitating the recent surge in rising expectations of the world's underprivileged masses, as characterized in Max Way's powerful essay, *The Dynamite of Rising Expectations*. He writes:

> . . . .a dangerous chemistry at work in the world today, that begins with the ferment of hope and can end in explosions of wrath, despite—or more likely because of—this nation's capacity to generate technical progress. Hope hardens into expectancy which soon turns to overexpectancy, then follows frustration and rebellion. The smoldering expectancy of the underprivileged masses is fanned by increasing awareness of what technology can do, in fact has done, for other men.
>
> Expectational fever has many carriers. Recent U.S. civil strife can be attributed in part to the impact of television on underprivileged blacks who see another world in their living rooms. A young black antipoverty worker says: "What whitey doesn't understand is that America is a damn rich country and that she flashes this richness in the black's face and tells him 'It isn't for you'."
>
> The transistor radio has a much longer reach. It spreads expectations into Bedouin tents, into huts on the slopes of Kilimanjaro, into Pakistani villages, and into the homes of Andean Indians. A truly massive revolution in communications has put classes and nations within sight and sound of one another. These confrontations have the power of dissolving or weakening social patterns that have endured for centuries. Change begins to seem possible, then desirable, then mandatory.*

Awareness that men elsewhere have it much better is not the only factor behind rising expectations; knowing of man's vastly increased power-to-accomplish has stimulated the masses to expect more.

> Achievements in space have become a fulcrum for the lifting of expectations. If we can put a man in space why can't we: eradicate slums?, beautify America?, find a cure for cancer?, reduce juvenile delinquency?, improve the lot of the Navaho Indians?, end war?, get

* Max Ways, "The Dynamite of Rising Expectations," Fortune, May 1968.

> a decent laboratory in the high school? In a way, each of these is a "good question." Whatever else the space program did it surely was the greatest stimulator of expectations of all kinds. Suddenly resources seem limitless and instantly applicable to men's needs. When poverty was accepted for generations—and not even recognized as poverty—it becomes overnight a burden not to be borne, an exploitation, a wanton injustice.†

Of course, modern communication media keep the more-privileged as well as the less-privileged better informed about what goes on in the world. This greater awareness probably contributes heavily to the younger generation's diminished respect for certain institutions. Observe how much of what the young are told by books and elders about social justice and morality is inconsistent with the fact and fiction regularly appearing on movie and television screens. The visual testimony is indeed more persuasive than the words it conflicts with.

Institutional Effects

Among the least visible of technology's collective effects are the impacts on institutions, social, economic, and political, which will probably prove to be more troublesome than environmental and other readily recognized consequences. As a case study of the aggregative effects technology can have on an institution, the next several chapters are given to the challenges brought to government. The political system was selected for this purpose because the strains that technology imposes on it are particularly worth noting.

† Ibid.

AN ASIDE ON TERMINOLOGY

Appendix

Obviously *technology* is on the tip of many tongues and pens. But to many users, probably a majority, the term is as vague as it is familiar. However, some users do have something specific in mind when they use the word; unfortunately, there are different interpretations in circulation, some broader, some narrower, than the view adopted herein. Here are some samples.

In a broader interpretation, technology is viewed as the accumulated skills, know-how, and techniques with which a society provides its goods and services. In the case of agriculture this includes all of the machines and other contributions engineering has made *plus* all the practical procedural know-how that has accumulated over the centuries (i.e., plowing practices, crop rotation schemes, and all of the "art" associated with managing an agricultural enterprise). Persons who use technology in its broadest sense are referring to the totality of man's capabilities. I have no quarrel with this view; I simply find it less useful, particularly in an engineering book.

My interpretation of the term is narrower, but not a great deal. It emphasizes the material means—machines, structures, and other physical contrivances—that a society employs to provide its goods and services. (However, you will probably notice that those who state or imply a broader sense, when they get down to specifics, usually talk about power plants, automobiles, computers, oil refineries, and the like—physical contrivances, in other words).

There are more restricted interpretations than that employed in this book. One you might encounter is the expression "engineering technology," or technology, employed in referring to engineering know-how—the practical (versus theoretical) side of engineering. Educational programs under these labels stress the more practical content of a typical engineering curriculum.

I am trying to make the best of an unfortunate terminology situation. The word technology is too much a part of everyday vocabulary; I can hardly afford to ignore it. But to use it raises the question, "In what sense?" The loose usage of news media and hallway conversation are unacceptable in a textbook, yet *any* specific interpretation makes some readers unhappy.

EXERCISES

1. Research and write an "impact case history," dwelling mainly on the technical, economic, social and other effects of a development such as the waterwheel, brick, self-starter for automobiles, cannon, electric motor, steam engine, icebox, or similar contrivance.

2. Identify engineering creations that have affected the farmer, housewife, or factory worker. Comment on the effects of those contrivances in aggregate.

3. It has probably never occurred to you that the unheralded invention of the wide-flanged ("I") steel beam has had some very significant effects. Please pass along the results of your research and reflections on the matter.

4. In Chapters 2, 4, 6, 7, 8, and 15 you have learned something about professional responsibilities of engineers. Tie all these comments together and add your own thoughts on the matter to produce a "statement on engineering professionalism." Don't overlook "after five" (off the job) responsibilities and opportunities.

5. So says J.J.G. McCue: "Engineers seldom look upon themselves as inciters and fomenters of revolution. But a revolution need not be bloody; it is any drastic transformation of social institutions. For many years now, one of the principal sources of pressure on existing social institutions has been engineering." But McCue is only one among many who are saying the same thing of engineers. Social revolutionaries they call us. What do they mean?

TECHNOLOGY AND GOVERNMENT

Chapter 16

To appreciate the interaction between technology and government, you should be alert to a variety of relatively recent techno-political changes. Think for a moment about the total consequence of these developments: jet aircraft, computers, atomic bombs, television, rocket engines for space flight, intercontinental ballistic missiles, mechanical cotton pickers, and tape-controlled production equipment. Far reaching, to be sure, all blossoming during or since World War II. In aggregate these and numerous other post-1940 developments are significantly different, both in nature and in consequence, from pre-1940 products of engineering. If you view the long-term trends in the pace with which new capabilities are developed and applied, the power embodied in them, the scale of these endeavors, and other important qualities, you can't help concluding that there is a major discontinuity around the World War II period.

What Has Changed?

In general, the power for good and for evil embodied in post-1940 developments is greater. In the case of nuclear weapons, this is conspicuous, but don't underestimate the power of television to influence attitudes.

But power is not the only difference; mankind has obviously become more ambitious in its engineering feats, in terms of scale, venturesomeness, funds and other resources required, and complexity. Particularly consequential is the expanded scale of these undertakings, such as man's probes into space, modern jet transports, our interstate highway system, and numerous other private and public ventures.

Furthermore, these days there are many more people around to be affected by what engineers create. In addition, their distribution is far from uniform; now there are enormous concentrations of humanity that amplify the effects of technology. And, too, people who live in a city are highly dependent on technical systems for most essentials of life. You can imagine the problems that breakdowns of water supply and sewage disposal systems would create, and you know what happens when a major power network cuts out.

Another change is that the political-economic climate for science and engineering has become much more favorable since 1940. The enormous post-World War II federal budgets for scientific and engineering undertakings are manifestations of this increased enthusiasm. Over this period the climate has been so favorable that, for some ventures, technical feasibility has seemed sufficient justification when social desirability would have dictated otherwise.

"The greatest discovery of the techno-scientific community since World War II is the U.S. Treasury."

Something else that is different is the pace with which new technologies are developed and put to widespread use. The development programs of some new technical capabilities are of the "forced-draft," if not crash, type. Furthermore, in days past, years or more likely decades elapsed between an invention's conception and the point where significant numbers of people felt its effects. Today we are prone to rush into widespread application.

What all this amounts to is accelerated development and application of a more powerful technology in a world of increasing population density. One major upshot is that government and governing will never be the same. Modern technology has brought added opportunities, new strains, and different circumstances in a variety of other respects to all levels of government. You should know something about these effects, so here is a sampling.

Technology's Impact on Government

Demanding Decisions

There has been a major change in the types of decisions made by public officials. Now they must make many technical decisions involving the bulk of taxpayers' money. These decisions concern dams, highways, supersonic aircraft, ballistic missile systems, mass transportation, computers, water supplies, energy conversion, pollution abatement, traffic control, weather satellites, and you can continue the list.

For one thing, engineering continually enlarges the number of alternatives. Take the generation of electric power; within a few decades the number of conversion methods has tripled. In addition to several nuclear methods, we now know how to generate power by exotic-sounding processes such as magnetohydrodynamics, thermionic conversion, and thermoelectric generation.

Actually, when it comes to technical achievements, there isn't much that man can't do. We can transport men to Mars, to . . .; we can supply ample water for everyone; we can satisfactorily dispose of the old auto hulks, disposable containers and, in fact, all human wastes; we can make automobiles a lot safer and the skies a lot cleaner; we can build planes that will travel as fast as you will want to go; we can influence the weather. The possibilities are almost unlimited. All we need is someone to pay the bill—and the price tags on many of these alternatives are staggering. The relevant question is no longer, "What *can* we do?" Now we ask "What *should* we do?" Obviously taxpayers are neither willing nor able to finance all of these, so priorities must be established and choices made.

This leads to an enormously complex decision-making task for public administrators and legislative bodies, especially Congress, holder of the national purse strings. It must allocate a staggering sum of money to a limited number of undertakings, limited because however large you may think that sum is, it can only support a fraction of the opportunities available. So public officials must choose. The matter is complicated further by the fact that a majority of the choices in one way or another involve science or engineering. Imagine yourself a Congressman reviewing budget proposals submitted by various departments of the federal government. You are zeroing in on one narrow facet of the total budgetary picture, wrestling with questions such as: "How much can we justify spending on transportation?" "Of that amount, how much should be allocated to development of new capabilities, how much to application of existing capabilities?" "Of the money to be spent on developing new transportation modes, how much should go to space travel, how much to air transportation, how much to surface systems, and how much to subsurface modes?" (You are probably picturing how complicated this becomes. If not, Figure 16-1 will help.) Of course, in competition with transportation for funds are all other technical

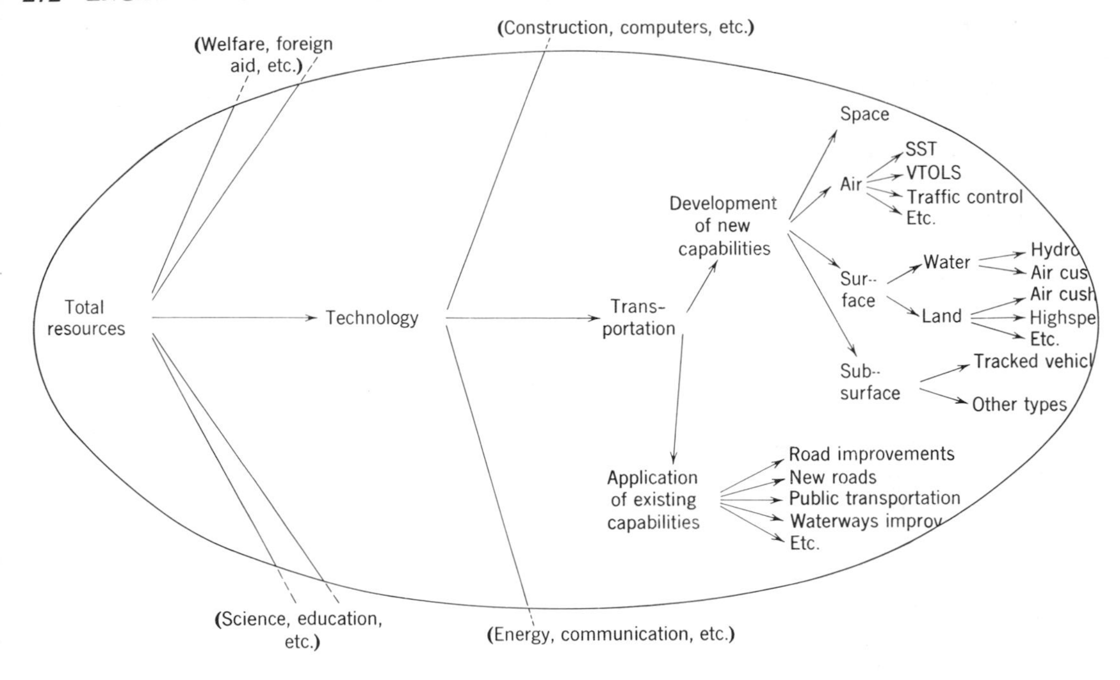

Figure 16-1 *What resources should go to what alternatives? What opportunities should be foregone or postponed? Resources are limited. If this diagram were completed, it would require a space the size of a gym floor. It would, indeed, reflect the magnitude of the allocation problem confronting federal officials. The task is a staggering one partly because of the number of alternatives; everything seems to explode into numerous subalternatives, yielding thousands of choices.*

opportunities, scientific research, welfare, foreign aid, education, maritime subsidies, U.S. Forest Service, . . . you know how the federal budget goes on and on.

But the number of alternatives is not the only problem; the complexity of technical alternatives continues to compound. Consider some of those new methods of generating electricity; their names are a challenge, to say nothing of their technical intricacies. What this generally means for decision makers is typified by the consequences for military planners and the legislators who try to evaluate military budget requests. Their decisions involve sophisticated weapons systems, large computer complexes, surveillance networks consisting of radar, satellites, and special aircraft, elaborate communication systems, huge construction projects, and so forth.

Furthermore, the opportunities for and consequences of blunders in making decisions of this kind have reached alarming proportions, for reasons cited at the start of this chapter. Nuclear energy *does* offer an unprecedented potential for harm; there are several ways we might unwittingly and significantly reduce the ozone content of the upper atmosphere which, in turn, will diminish its important capacity to shield us from ultraviolet radiation.

All this has significantly altered the life of the legislator and public administrator; ask one. He has at least as many "people problems" as his predecessor of 50 years ago, plus the added burden of many technical problems.

The Founding Fathers Could Not Have Envisioned It

Who would take the trouble to count them, the federal government is far too complex, but the list of federal departments, commissions, boards, agencies, and administrations that owe their existence to a technical capability goes on for pages. It includes some familiar names: National Aeronautics and Space Administration, Department of Transportation, Federal Power Commission, Civil Aeronautics Board, and the Federal Communications Commission. The functions of these agencies are to regulate or promote a technical capability. In doing this, they exert enormous influence over our daily lives.

Some of them, mainly the "superagencies" like NASA and the Department of Defense (the nation's number-one developer and user of technology), have some unusual qualities. One is an organizational expertise that enables them to bring vast brainpower and financial resources to bear on problems of staggering complexity and to succeed. The Apollo program is a conspicuous example but, for every Apollo, there are dozens of "superprojects" that receive no such publicity. Development and manufacture of the Polaris submarine involved 11,000 subcontractors, giving you an idea of the complexity of such an undertaking.

These technopolitical institutions are powerful forces in both a technical and a political sense, no likes of which were around in 1789. And so, for Congressmen and top administrators, the use and control of these organizations are relatively new challenges—challenges that may be greater than you think, since these organizations attend to their lobbying and public relations in promoting the particular technical capability to which they are dedicated.

Local Government

State and local officials will be quick to remind you that it is not only the federal government that is affected. You can imagine what it means to city officials to find industries emigrating and large numbers of unemployed persons immigrating. Of course, both trends owe much to machines of one kind or another. Furthermore, picture the headaches involved in:

- **Deciding on the wisdom of a new connector between center city and the circumferential freeway.**
- **Locating such a roadway, especially when every alternative takes it through a densely populated area.**
- **Choosing between such a roadway and a public transportation system.**
- **Siting a new airport.**
- **Trying to alleviate interminable traffic and parking problems.**
- **Deciding what to do when the city's solid-waste incinerators no longer meet state-legislated pollution standards.**
- **Passing a new building code proposal that involves all kinds of technical standards.**
- **Deciding whether to install a computer that would provide a master information system for all city departments.**
- **Determining the economic wisdom of installing a system that converts sewage sludge into electricity.**

The technical challenge to them is the combined result of technology-precipitated problems (e.g., pollution) and of new technological opportunities (e.g., computerized record keeping).

Technology and Government—The Interaction

Modern technology affects government in many ways. But the reverse is also true; government has a big hand in determining the nature and role of technology. This influence is worth examining.

Government's Influence Over Technology

Government can't help influencing technology as long as it is by far the largest financial backer. The federal government buys weapons, trucks, computers, meterological equipment, and so forth, the cost of which consumes a sizable chunk of the federal budget. In spite of the vast sums and therefore the influence involved, this is not the major cause for concern. More consequential is government's role in making the following types of decisions.

Type I Decisions

If man acquires the ability to get to Mars, achieves a breakthrough in his ability to tunnel beneath the earth's surface, develops a practical means of generating electricity through nuclear fusion, or acquires almost any major new technical capability, it will be because he has *chosen* to do so. Today very few of these capabilities are simply "suddenly discovered." Their evolution is probably much less accidental and much more deliberate than you imagine. Ordinarily, a new capability is the result of a goal-oriented process that begins with a decision to commit financial and manpower resources to the development of that capability (e.g., power from the atom). This is followed by a period of intense effort spanning months or probably years, usually resulting in the sought-after capability. *Therefore, much of the new technology that evolves is planned, meaning that there is an opportunity to choose, and meaning that this is the logical time to choke off technical ventures that will ultimately cause society more grief than benefit.*

Consider the prospect of mass-produceable, artificial, implantable hearts. It's an appealing thought: the extension of human lives by years through the replacement of defective hearts. But this may not be such a big favor to mankind at that. With it come some knotty social, legal, and moral headaches. Overpopulation is already an enormous problem. And if attempts were made to restrict the use of such devices to only the "needy" cases, where would you draw the line and who would rule in specific cases?

There is little doubt that such devices can be developed. True, engineers and medical researchers lack some of the necessary know-how at the moment, but this can be acquired. What's lacking is the decision to proceed and the funds. (What is your recommendation in this case: proceed or not?) The issue is whether or not to develop a *new* capability, referred to henceforth as a type I decision.

Type II Decisions

We *have* the ability to continue sending explorers to the moon, to dig a sea-level canal across Panama, to build a new freeway through any of our major cities, to provide artificial kidneys to all in need, to factory-build a sufficient number of moderately priced housing units, or . . . . *Should* we (not *can* we) do so? These decisions, concerning the application of *existing* capabilities, are referred to here as type II decisions.

Here is one. It is now possible to maintain a central, computer-kept file that merges the previously separate files of social agencies, tax bureaus, police departments and other government agencies. A multiple-terminal time-sharing system makes this "data bank" possible. By eliminating expensive duplication of files, a national, state, or urban data bank can save taxpayers a lot of money. It can also improve the effectiveness of the sharing agencies; all law enforcement agencies, for example, can pool their information. In addition, such a comprehensive accumulation of data could be processed to provide invaluable statistics to social scientists and government planners. These and other benefits make the data bank idea attractive in some quarters.

But there is a risk involved: the possibility that such a system will be used to violate the privacy of citizens. It is conceivable that unauthorized persons or agencies could obtain confidential information. This and other potential abuses of citizens by this electronic "big brother" have stirred opposition, Congressional concern, and writers for the Sunday newspaper supplements.

This decision concerns the use of an *existing* capability; the computer equipment necessary for a national data bank has been around for awhile. Should government use this capability?

Who Makes These Decisions?

Who is it that makes the type I decisions and thereby determines what technical possibilities become capabilities? Who makes the type II decisions and thereby determines how and where our technology is employed? What forces are at work in influencing these choices?

Government, business, and the public all have a say in these matters. Although the exact role of each is impossible to pinpoint, observers have no problem detecting major shifts in relative influence. They can tell you with confidence that:

- ***Over the past few decades, government has assumed the major role as type I decision maker.*** **It now makes most important decisions of this kind. The federal government now spends billions of dollars on the development of new engineering know-how that spans national defense, transportation, space exploration, atomic energy—you name it. Public funds allocated to research and development have increased dramatically since 1940. The prime significance of this deep involvement is that the federal government has become the prime determiner of the new technologies that man acquires.**

Could workable, implantable artificial hearts become available without the federal government's backing? Possible, but not probable. The VTOLS described earlier would be delayed years, probably decades, if their development depended on private financing and the demands of the consumer market. This shift in influence from private enterprise and the marketplace to government is significant, indeed. This is one of those major changes that schoolbooks haven't caught up with. And how about you, are you still crediting private enterprise for significant developments that have evolved from government-financed R&D programs? You would probably be surprised to learn just how much of the digital computer's capabilities was paid for by the federal government. Time sharing, graphic displays, light pens, the core memory (on which computers depend so highly) and other hardware owe their existence to the "federal financial funnel." A similar story applies to jet aircraft.

- ***Concerning type II decisions—government and big business now exert more influence on how and where technology is applied (as opposed to developed), the public less, than you would like to think.*** **Most informed observers seem pessimistic about what effects the wishes of the citizenry have on where and how government uses its bulldozers, rockets, computers, ... (This is not to imply that this situation can be *entirely* otherwise in view of the complexity of these undertakings and of the issues involved.)**

But, you say, surely consumers, by expressing their preferences at the marketplace, influence what big business does with its technical capabilities. Yet here, too, you may be naive. Today, mass persuasion techniques make grass roots consumer preferences almost a thing of the past. Consider the power of television in this respect. Do you believe that the auto industry has marketed 400-horsepower automobiles in response to demands by consumers? If so, you underestimate the behavior-influencing techniques of "Madison Avenue." As a consequence, it is impossible to say what influence unadulterated consumer preferences have, because consumer behavior at the marketplace is so conditioned by the promotional process. Thus it appears that Everyman has less say than he would like—as citizen *and* as consumer—about the manner in which modern technology affects him.

While government assumes an expanding role in making type I and type II decisions, these highly consequential decisions fall to a disturbingly small number of people.

A New Decision-Making Elite

A substantial proportion of the federal budget is now allocated to scientific and technical ventures that are unknown or at best mystifying to the citizenry. As time passes, more and more multibillion dollar decisions are being made by a relative handful of people, while the electorate becomes more and more out of touch with the possibilities and considerations. The result is a new elite consisting of persons in key government positions, *plus* their advisors in and out of government, who are making these highly consequential decisions. These men have acquired unusual influence as a result of their access to technical and scientific knowledge that the masses no longer comprehend.

This concentration of decision-making responsibility in the absence of public consensus—even awareness—is hardly consistent with what the founders of this nation visualized. It is a serious matter that causes an increasing number of thinkers to wonder if our form of government is being slowly undermined because the electorate is unable to make intelligent evaluations in this age of complex technology. Admiral Hyman Rickover discusses this matter in his brilliant essay, "A Humanistic Technology."

> The Founding Fathers were well aware that democracy is the most difficult form of government. They knew that to make a success of it, a people must have political sagacity as well as what the ancients called "public virtues"—a combination of independence, self-

reliance and readiness to assume civic responsibilities. But they felt that Americans possessed these qualities; that, indeed, the conditions of life in America developed just the type of man who would know how to make democracy work.

Among the advantages favoring a workable democracy, the founders counted the fact that Americans were for the most part independent farmers, artisans and merchants. Being used to managing their own business, such men would, they felt, know how to manage the nation. A scarce population and the immense wealth of the country in land and other resources would prevent formation of a propertyless class dependent on others for employment. The political equality basic to our system of government would thus be firmly supported by real equality among our people. The founders were convinced there would be free land for generations to come; they could not have envisioned a hundred-fold population increase in but two centuries. That seventy per cent of our people now live in urban conglomerations would have horrified them; they judged Europe's propertyless urban masses unfit to govern themselves! To them America's unique advantages were a guarantee of success for their political experiment. They felt that the land, the people and the political system were made for each other.

These special advantages are nearly all gone now. They began to disappear with the coming to our shores of the Industrial Revolution roughly a century ago; we are losing them at an accelerated rate since the full impact of the Scientific Revolution hit us about two decades ago. Directly or indirectly it has been the new technology these revolutions brought into being that altered the pattern of national life in ways that are detrimental to the democratic process. The many benefits we gain through technology come at a cost.

Though we save ourselves much unpleasant labor by means of technology, we have to exert ourselves more than in the past to reach the competencies required of all who are involved with technology. Decision makers now should have a liberal education as well as professional competence, so must workers have a basic education in addition to their specific vocational skill. This is the price we have to pay for the many good things technology can provide.

But the raising of educational levels is not limited to job requirements. It is also essential to the discharge of our responsibilities as democratic citizens. Where in the past, life itself developed in most Americans the wisdom and experience they needed to reach intelligent opinion on public issues and to choose wisely among candidates for public office, we must today acquire this competence largely through studies that many people do not find particularly congenial. Yet unless one understands the world he lives in, including technical issues requiring political solutions, he is not a productive, contributing member of society. Uneducated citizens are potentially as dangerous to the proper functioning of our democratic institutions as are uneducated workers when they handle complicated machinery.

By making it possible for affluence and leisure to be spread over large segments of the population—theoretically over all the people —*technology gives support to democratic institutions*. We are approaching a situation comparable to that of Athens 2,500 years ago where every citizen was an active participant in the governance of his city state. He would not have had the leisure to do this, had there not been slavery. Today each of us has many more mechanical slaves than the Athenian had live ones. We have at least as much leisure as he had to devote to public affairs.

The Athenian, however, dealt with public issues that were not beyond the comprehension of citizens. Modern democratic citizens, on the other hand, are faced with issues that are extremely difficult for laymen to understand. They must depend, to an extraordinary degree, on the advice of experts. Whether such advice is competent, as well as impartial, is often hard to judge. Much of the difficulty arises from the complexities technology introduces into modern life. To the extent that it renders public issues incomprehensible to ordinary citizens, *technology undermines democratic institutions*. It sets a lower limit to the educated intelligence citizens must have if they are to meet their public responsibilities—a sort of Plimsoll mark. Those who fall below this mark are precluded from participating in the public dialogue through which consensus is formed in free societies; they are precluded simply because they do not understand public issues involving technology. Democracy is not viable if too many fall below this mark.*

EXERCISES

1. Recall the type I decision concerning implantable artificial hearts. What is your decision? Cite your reasoning.

2. Consider the sample type II decision concerning a national data bank. What is your recommendation? State your case.

* Excerpts from a Granada Lecture before the British Association for the Advancement of Science, by Vice-Admiral Hyman G. Rickover, 1965.

3. Some observers have reflected that Adolph Hitler's main impact has been *since* World War II. What do you suppose they mean?

4. With the help of recent newspapers and news weeklies, identify examples of type I and type II decisions currently under debate.

5. Focusing on energy, cite examples of type I and type II decisions that have been or are being made by public officials. What major decisions concerning energy sources, generation, and utilization are left for decision makers in private industry?

HOW WE CHOOSE

Chapter 17

*All knowledge is good knowledge; not all application is good application. Unless our political servants avoid the cowardice of a too-limited application of technology, unless they avoid the moral turpitude of mis-application, unless they avoid the immoral squandermania which results from a lack of stringency in scheduling proximate goals, then, indeed, the advances of technology could be catastrophically disastrous.**

Since the federal government has become the major type I decision maker and, therefore, the main determiner of the directions in which technology is expanded and the pace at which this new know-how is acquired, and since government now has a big hand in making type II decisions and, therefore, is a major determiner of how these capabilities are to be applied, our elected representatives face some important new challenges their predecessors never knew. They have the task of allocating dollar, brainpower, and other resources to the development and application of a substantial proportion of this nation's technical capabilities. And a crucial task it is, because there is enormous power here, for good and for ill. Whether technology works for the benefit of mankind or proves to be its undoing depends on the intelligence and diligence with which officials direct its evolution and application. With so much at stake in these decisions, it is important that you have a closer look at the manner in which they are made, particularly at the adequacy of the methods and competency of the people involved.

As you learn more about the origins of government's large-scale technical ventures, you will conclude, as many others have, that the process for allocating resources to such enterprises fails to recognize national goals and priorities adequately; that it considers too few of the alternative means of achieving these goals; that it is crude in its consideration of long-term consequences of technical ventures; and that it is too responsive to pressures from organizations with vested interests.

Who should know better about weaknesses in this allocation process than public officials? Congressmen, for instance, freely admit that they are unhappy with the manner in which they make type I and type II decisions. Congressman Ryan writes:

> A large, rich nation should engage in ambitious technological ventures. However, we should have the wisdom and the courage to admit

* Robert Watson-Watt, "Technology in the Modern World," *Technology and Culture,* 1962.

> that our efforts in this direction are far from ideal, that we have wasted resources and talents and have all too often misdirected the creative technological efforts on which we pride ourselves.
>
> I feel that there is room and need for great improvement. Congress and the Executive Branch have much to learn about rational decision-making based on the real costs and benefits of technical ventures.*

Another says:

> I believe the time is due for a comprehensive review of our national planning and decision-making system in the area of science and technology. Frankly, I do not believe Congress can devote enough time and resources to study a problem of this magnitude.†

These and other persons in government observe firsthand how public funds are allocated to technical alternatives; they are unhappy with what they see.

One of the first tangible signs of official awakening to inadequacies in this decision-making process was the appointment by Congress of a special fact-finding Commission on Technology, Automation, and Economic Progress. Some of this commission's concerns are apparent in these words from its report.

> There are two questions before us. One is: What proportions of national income should go for what ends? The second is: How adequate are our mechanisms for making such decisions and assessing the consequences?
>
> There is a widely held belief—derived from our experience in military and space technology—that few tasks are beyond our technological capability if we concentrate enough money and manpower upon them. However our relative affluence and technical sophistication should not lead us to believe that we can attain all our goals at once. Their cost still exceeds our resources. Thus we will continue to face the need to set priorities and to make choices.
>
> We cannot set forth in this report what the specific priorities should

* Congressman William F. Ryan, "Apollo and the Decay of Technical Excellence," *The American Engineer,* January 1968.

† Representative Joseph E. Karth, "The Congress, Public Policy and Science," *TRW Space Log,* Spring 1966.

> be. We are concerned with *how* we decide what to choose. Congress has asked us: "How can human and community needs be met?" But there is a prior question: "How can they be more readily recognized and agreed upon?"
>
> What concerns us is that we have no such ready means for agreement, that such decisions are often made piecemeal with no relation to each other, that vested interests are often able to obtain unjust shares, and that few mechanisms are available which allow us to see the range of alternatives and thus enable us to choose with a comprehension of the consequences of our choices.*

Obviously the Commission feels, and many members of Congress and administrative agencies have so commented, that the absence of specific national goals and associated priorities is a serious handicap. Of course, we have goals, generally implicit, that have evolved over the years (e.g., security from aggression), but nothing resembling the deliberately established, comprehensive, operational set of goals that administrators and legislators should have. They deserve some sympathy; it *is* difficult to allocate resources intelligently as long as goals and priorities are so elusive. But this is not the only deficiency.

Regrettably, in the allocation process, insufficient consideration is given to alternative courses of action. Too often, for instance, a decision is made to underwrite the development of a new capability without giving a fair hearing to other ways of accomplishing the same objective *and* other entirely different uses of the money. We have spent billions in public funds trying to develop an economical means of converting nuclear energy into electrical power. We have succeeded, but promising alternative sources of power have been known for some time, and they have received little serious consideration and negligible public funds for development. This is a familiar pattern. In many instances, this happens because one alternative has a champion—a federal agency, an industry, or a pork-barrel-conscious legislator—giving it an inside track over all others, regardless of merits and costs to society.

Furthermore, there is no satisfactory mechanism at present for assessing the total costs and benefits of alternatives and their relative contributions to national goals. Congressmen have no

* *Technology and the American Economy,* Report of the National Commission on Technology, Automation, and Economic Progress, Volume 1, February 1966.

adequate source of answers to questions like: "What would be the long-run costs of the SST?" "The benefits?" "The indirect effects?" "The risks?" "How does it compare with alternatives?" The SST was chosen to illustrate this point because, during the 1971 Congressional deliberations that resulted in termination of the United States SST project, we saw a scandalous display of ignorance, bias, distortion, lobbying, emotion, unsubstantiated claims, media-fanned scare stories, conflicting testimony, and just about everything else you would *not* want in or surrounding such deliberations. This is what happens when the stakes are so high, the issues so burning, and the facts so sparse. What a way to decide!

In the preceding chapter you were confronted with a sample type I question about artificial hearts and a typical type II question about a national data bank. You are not equipped to answer those questions until you know, specifically, the benefits expected from each venture, total costs anticipated over the long run, their indirect effects, and alternative uses of the resources required. In fact, before a commitment is made, you should also know what kind of difficulties to expect in controlling the use of artificial hearts and abuse of a national data bank. Similarly, unless public officials and business executives have this same kind of information, they too are ill-equipped to make such decisions. But they do it all the time. Certainly decision makers are entitled to some mistakes, even an occasional blunder; these *are* very complex decisions. But we *do* have a right to expect them to ask the right questions, to obtain better information before making technical decisions affecting millions of people and requiring major commitments of resources.

The People Involved

How qualified are Congressmen and other public officials to make such decisions? In Congress, for instance, the number of legislators with a technical education has been running at about a half dozen. In other branches the proportion is not much different. So you can hardly say public officials, as a group, have a background commensurate with technically demanding decisions. By their own admission, they are ill-equipped.

So, you say, if most officials are not qualified, and if comprehensive, objective information is not readily available, why not call in "experts" and rely on their advice? Congress does rely heavily on expert testimony, but it turns out to be a prime source of disenchantment with the whole decision-making process. Most experts who come before Congress to offer testimony on costs, benefits, and risks associated with some technical or scientific venture are biased. Representatives of federal agencies usually provide overly optimistic appraisals of their proposed programs at Congressional budgetary hearings.

Well, you say, why not rely on experts outside government to provide Congress with appraisals of technical ventures proposed by administrative agencies? Congress does this, but these advisers have their biases, too; they are discipline-oriented if not agency-oriented. Biologists will testify in favor of almost anything that will improve the environment; authorities on urban affairs feel that not enough money can go to the cities; aerospace people are appalled that we are not spending more on space exploration; and so it goes.

Here is an illustration. Not long ago a proposal to drill through the Earth's crust to the mantle, known as the Mohole Project, was the object of much discussion in Congressional hearings. Over the course of prolonged testimony, there were references to Russian efforts to do likewise, creating the impression that the two nations were involved in a "race to the mantle." Yet there was no evidence to confirm this and, since then, observers have concluded that the supporters of this project let their enthusiasm get the best of them while they were "advising" Congress.

The elaborate technical advisory system now extensively employed by the Executive and Legislative branches of the federal government is vulnerable to criticism, and has gotten it.* The upshot of these agency, mission, and discipline biases, after reams of testimony preparatory to a technical decision, is that officials must do a lot of second-guessing.

Of course, inviting a progression of experts to testify on the merits of a proposed engineering venture opens the door to spokesmen for special interests.

Unsolicited "Advice"; Lobbies; Industrial Constituencies

Vested interest groups, lobbying, and pressure politics are not new to you. But you may not be aware of the amount of pressure

* Daniel Greenberg. "Don't Ask Your Barber Whether You Need A Haircut," *Saturday Review,* November 25, 1972. This article is the kind that makes you stop and think, and even chuckle. But the one by Martin Perl in *Science,* September 24, 1971, entitled "The Scientific Advisory System: Some Observations," causes you to stop and worry.

that government agencies themselves exert on legislators to sway them in favor of pet projects. The power plays, politicking, the infighting—you wouldn't believe what goes on.

Each federal agency has a powerful source of pressure in the research institutions and manufacturing organizations that derive large financial benefits from the agency's activities and programs. This is a source of dismay to many citizens. Nowadays, an agency simply hints that one of its proposed programs may not get the financial support requested from Congress, and its followers will set their lobbying machinery in motion. This news item appearing in *Science* describes a brief episode in a budget meeting between National Science Foundation officials and a Congressional committee.

> Senator Gordon Allott took the extraordinary step of warning NSF to refrain from repeating the lobbying campaign that helped retrieve part of the Foundation's budget last year. Though there was no evidence that NSF directly inspired that campaign, Allott charged that NSF turned its clients loose on the U. S. Congress. "I would say every Senator in the United States Senate was absolutely besieged and lobbied by College people in his own state . . . ," Allott declared. "I believe that the members of the NSF may have utilized monies of the Federal government in contacting various institutions throughout the country. . . . If we have a recurrence of a situation like that again, I assure you that there is going to be a long and detailed investigation into the situation." *No discussion was held of why NSF should be barred from even discreetly playing a game that is blatantly engaged in by virtually every other federal agency.**

Much has been written in the recent past about the followings that technical superagencies have cultivated. The most widely quoted commentary was made by President Eisenhower. He was warning the nation—and he was in a position to know—of the existence and dangers of a military-industrial complex which, in his words, is a "conjunction of an immense military establishment and a large arms industry," exerting an influence "felt in every city, every state house, and every office of the federal government." He called on the nation "to guard against the acquisition of unwarranted influence."

The primary significance of these powerful constituencies of federal agencies is that they have much at stake in certain prop-

* *Science,* May 3, 1968. My italics.

osed engineering projects. When such a proposal is under discussion in Congress, they throw their weight behind the sponsoring agency. When multibillion dollar contracts are at stake, you can be sure there is some sweet and fast talk in the smoke-filled back rooms. Unfortunately, there is no counterbalancing pressure representing the best interests of society.

Thus, some of the handicaps Congressmen and other allocators of public funds must work under are: their own nontechnical backgrounds; inadequate and often misleading advice; unsatisfactory benefit-cost appraisals; and pressures from vested-interest groups. The result is that we have been led down some unprofitable, sometimes hazardous, technological paths. But let's not be too hard on the legislators; the supporting systems (e.g., the information resources and appraisal mechanisms) are more at fault than the individuals. That system is geared more for the relatively simple, slow-moving, sparsely populated world of two centuries ago, when man's power to control nature and other men was insignificant compared to now.

So much for the bad news, the limitations of the means with which this nation makes its major techno-political decisions. There is cause for concern. Yet there is some partially compensating good news; Chapter 18 describes it.

EXERCISES

1. For a revealing experience, take a major technical matter that Congress has had to cope with and research *some* (to do a comprehensive job could take months) of the debate, testimony, and issues associated with the decision. Possibilities: SST, ABM, Occupation Safety and Health Act, Environmental Policy Act, Alaskan pipeline, space shuttle, or automotive emission control.

2. Prepare a report entitled "A Contemporary Technical Controversy—Facts and Views." Scan periodicals to locate articles and editorials on a "hot" technical topic such as a proposed nuclear power plant, highway, pollution controversy, pipeline, weapon system, energy matter, waste management issue, or piece of legislation. Prepare an abstract of each piece that you select *and* your commentary which, among other things, should point to apparent bias, distortion, disagreement, and contradictions. Your selections should not be restricted to one point of view or a particular point in time.

3. An oil company proposes to build a tanker terminal on the ocean approximately 2 miles offshore. The oil will be transported to land by an underwater pipeline. A special committee of state legislators is holding an open hearing with the prospect of passing legislation that will regulate or ban such a terminal. Represented at the hearing are: the oil companies, local environmentalists, Environmental Defense Fund, the Environmental Protection Agency, Department of Defense, the President's energy advisor, Federation for Fossil Fuels, seashore resort commercial interests, officials of shore communities, and local union officials.

This is a role-playing exercise; you will be assigned to represent one of these special interests or to serve as a member of the legislative committee. Students in the latter capacity should come prepared with a list of questions that they believe this committee should have answers to before it proposes legislation, if it is to do so intelligently.

This can be an enjoyable as well as revealing experience, but it won't be if participants simply talk off the top of their heads; some advance thought to the role you are playing is essential to success.

4. You are a staff assistant to the administration of a large city. You have been assigned to investigate major mass transit alternatives and to identify criteria (officials would probably say *factors*, but you recognize that functionally these are *criteria*). Officials want a capsule description of each type (surface track, monorail, air supported, etc.) *and* a comprehensive list of criteria that they should consider in selection of the final system. If you do your job well, you will have an impressive array of alternatives and a long list of criteria, which together indicate the complexity of the decision facing officials.

TECHNOLOGY ASSESSMENT

Chapter 18

What might be done to improve government's means of making type I and type II decisions? You could make some appropriate suggestions simply on the basis of reason and the contents of Chapter 17. In so doing, surely you would call for:

1. **A strengthened mechanism for establishing national goals and associated priorities. You do agree—it is difficult to make means choices when ends are so vague.**
2. **Public decision makers to consider more alternatives before committing resources to a particular one.**
3. **More persons with technical or scientific backgrounds in government, particularly as elected officials.**
4. **A mechanism that will provide officials with objective assessments of the full impact of proposed technical ventures (in other words, with the information on costs, benefits, risks, etc., that they have been lacking).**

There is no significant progress to report for the first three of these proposals, but for the last, there is. The good news is the recent evolution of an evaluation process referred to as technology assessment.

Technology Assessment

Many authorities are using this expression to refer to an impact analysis of a technical undertaking. Although there is general

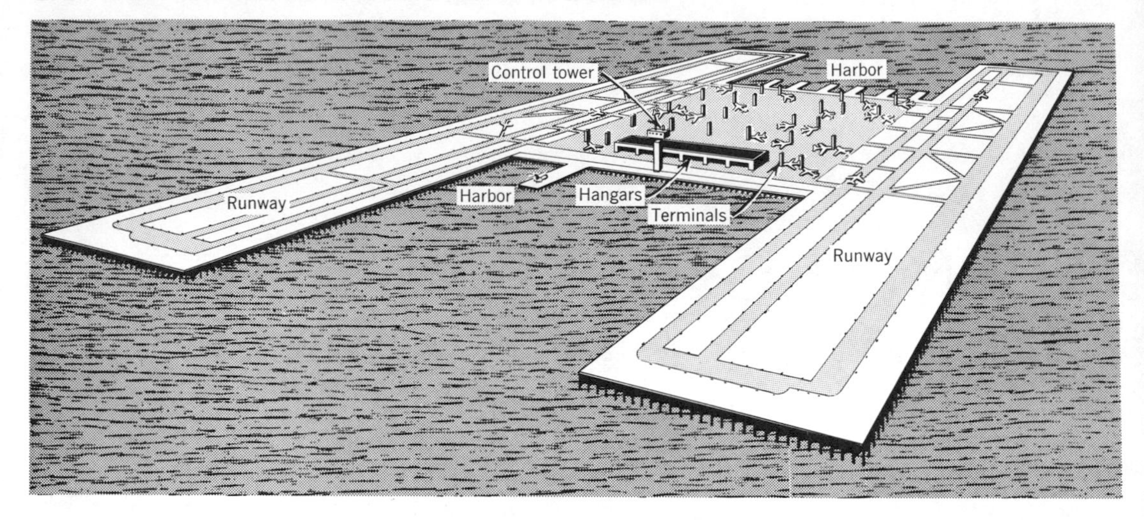

Figure 18-1 *This is a sketch of one of a surprising number of proposals for offshore jetports appearing in the literature in the past 10 years.*

agreement that technology assessment will serve man well, in this early stage, when it is still taking shape, there is no such agreement on what a technology assessment should consist of. So the following characterization of technology assessment, presented in a hypothetical illustration, is a composite based on the dialogues now taking place in the journals and meeting halls.

A jetport similar to that pictured in Figure 18-1 has been proposed for an offshore site not far from a large metropolitan area. This facility will be pile-supported, floating, land fill, or polder; the details have not been resolved. Before proceeding with this venture, a full appraisal should be made of its social, economic, and environmental effects and, according to this composite vision, technology assessment is exactly that. It calls for careful consideration of:

1. **The *appropriateness* of additional air transportation facilities. Is there a *need* for another airport in the area in view of projected numbers of travelers, of the possibility of new modes of transportation that would significantly affect the volume of jet traffic, etc.? There is a related matter, one that is rarely considered and is obviously more difficult to cope with. It has to do with the *wants* (desires if you prefer) of society. Strictly speaking, we don't *need* more jetports; we build them because a**

sufficient number of people presumably value speed and convenience more than the things (like a cleaner, quieter environment) that must be sacrificed to obtain those benefits. This is a trade-off that should be reevaluated each time expansion of the air travel system is considered, since it probably will not hold up indefinitely. To put it another way, the air travel interests will tell us that we *must* continue expanding air terminal facilities to accommodate future passenger volumes. Although it is rarely done now, eventually more people will challenge the assumption underlying this "expand indefinitely policy." Those who do are finally viewing what are usually cited as "needs" for what they really are: wants. These are complicated matters, true, but that hardly seems sufficient reason for neglecting them while type I and II decisions are made.

2. ***The alternatives;*** **if expansion of the city's air terminal facilities is appropriate, it would be foolish not to consider other means along with the offshore proposal.**
3. **The *benefits,* which in this example will be safer operations and less noise at the city's existing airport, assurance of adequate air travel facilities for some time to come, stimulus of the local economy, and so forth.**
4. **The *costs,* the obvious components of which are construction and operation. However, other less tangible costs are likely, many of which are environmental which, in turn, reduce the recreational value of the region.**
5. ***Other effects,*** **which cannot be identified in advance as positive, negative, or even as consequential. This facility will affect navigation and land use; it will probably affect land values; it could affect air and water currents. The magnitudes and the consequences of these effects can be predicted only after in-depth investigation—if then.**
6. ***Restraints necessary.*** **If restrictions on flight paths are forthcoming, if noise limits will eventually be imposed, if a minimum is desirable for the distance between jetport and shore, these should be established *before* the decision to build. These restraints most likely will affect the feasibility of the project.**

For introductory purposes certain features have been omitted from this hypothetical example that surely should be included in real assessments. For instance, the above case sounds like a decision in a vacuum. Yet technology assessments must transcend specific ventures because so many of these decisions are interrelated. Many technical undertakings affect the same atmosphere; many consume a given resource like oil; many can affect employment levels in a given area or occupation. If decisions concerning individual projects are made independently, government will be continuing its practice of creating new problems with its solutions to old ones. The interrelatedness of techno-political problems and decisions continues to increase. Surely you have observed how the so-called energy problem interacts with problems of economic growth, transportation, solid-waste management, environmental quality, balance of payments, and no doubt others. Yet interrelatedness is a characteristic of problems men are poorly equipped and seldom inclined to cope with. This surely is a prime challenge to technology assessors.

Let us hope that technology assessors will come to grips with the "conflicting constituency" problem. It's a headache to cope with and, consequently, tempting to slight. Comparing benefits with costs in the tire-mounter case is a relatively simple matter; comparing them when one segment of the population derives most or all of the benefits while another bears most or all of the costs is not at all so. Consider the bank that is installing a computer system. A primary benefit is improved service to bank customers; a major cost is the personal hardship suffered by bank employees displaced by the computer. The same kind of situation prevails in siting decisions for airports, highways, and power plants. Often there are more than two population segments that are affected differently and, to complicate the decision further, there is usually some overlap among these "conflicting constituencies." There is no easy out for decision makers on this score; this is another prime challenge to technology assessors. What we *can* expect, however, is that such constituency conflicts will be explicitly and realistically recognized in the assessment process.

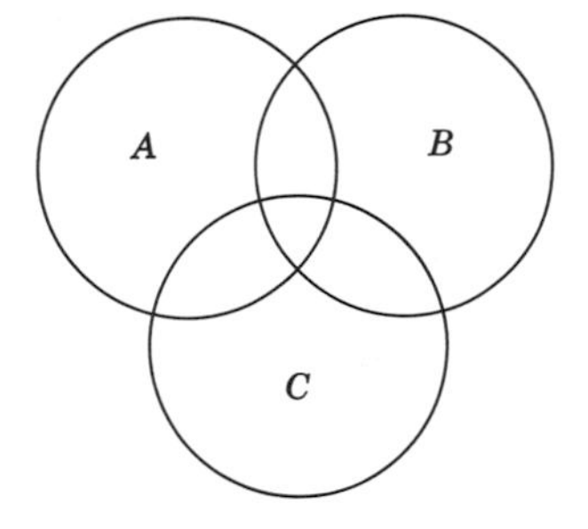

To be of maximum benefit, technology assessments must incorporate a "global" view; these decisions often produce consequences spanning many political boundaries. This aspect of assessments is likely to be slow in developing. After all, towns are still willing to discharge sewage for towns downstream to worry about. So what can we expect from nations?

Also, technology assessment need not—hopefully, it will not—be restricted to a reactionary role. Surely there are highly desirable capabilities that society has not benefited from because the economic incentives or the necessary institutional arrangements are lacking. Technology assessment rightfully can be expected to alert government to such opportunities.

Therefore, technology assessment is intended to supply much of the information that public officials need to make intelligent technical decisions. Although different authorities have somewhat different views of the character of technology assessment, most agree that, at a minimum, it should predict direct and indirect effects over the long run from the point of view of society at large. Note particularly the future orientation of technology assessment; it is primarily a means of *predicting* wants, costs, and benefits; of *anticipating* the need for regulatory legislation; of sounding *early warnings* of impending hazards. This feature plus the comprehensiveness of technology assessment seem to be its prime virtues.

Although the term and its comprehensiveness are new, many of the specific concepts and practices that comprise a technology assessment are not.

> Technology assessment is routinely carried out under a variety of labels, in both industry and government, in the pursuit of private and public objectives. The often-heard charge that ''technology is running wild'' fails to consider that uncounted possibilities for innovation are never translated into production or application precisely because the side effects are considered intolerable. Even, and perhaps especially, the much-criticized pharmaceuticals industry casts out large numbers of substances for every one it adopts. That no one keeps count does not mean that anything goes.
>
> The more recent idea of technology assessment as a responsibility of society, and thus of government, arises rather from the potential that modern science and technology afford for calamities on a very large scale, and because prior screening for undesirable consequences is often too costly for anyone but government.*

A close relative and recent forerunner of technology assessment is the provision of the 1969 National Environmental Policy Act requiring environmental impact statements from all federal agencies. That Act, incidentally, requires federal agencies to file a statement with the Environmental Protection Agency whenever a technical venture is contemplated. Thus, before the Corps of

* From *Resources,* January 1973, published by Resources for the Future, Inc., 1755 Massachusetts Avenue N. W., Washington D.C. 20036.

Engineers can start construction of a new dam, or before a proposed highway can be approved for federal funding by the Department of Transportation, a comprehensive statement must be submitted, covering environmental effects, irreversible commitments of resources, and alternatives to the proposed project. Observe that what the Environmental Policy Act requires for environmental impact, technology assessment provides for *total* impact.

Although the term and the concept of technology assessment originated in the late 1960s, by the early 1970s it was already widely known and discussed in political, scientific, and technical circles. The speed and extent to which the notion has spread indicates the widespread concern for the problem it is a response to, and the hope that many observers see in this approach. In fact, Congress recently voted itself a modest technology assessment capability in the form of an adjunct Office of Technology Assessment. A significant and welcome development indeed. Now members of Congress have an agency that has no specific technology to promote, staffed by professionals whose business it is to make independent appraisals of proposed technical ventures, to which they can turn for the information they need.

The appropriateness of Congress's action in creating its Office of Technology Assessment becomes more apparent as you learn more about the relative roles and resources of the Executive and Legislative branches in the techno-political decision-making process. Projects like the supersonic transport, the breeder reactor, and military weapons systems originate with agencies of the Executive branch and are sent to Congress for authorization of funding. Congress rarely receives a satisfactory explanation of the alternative programs considered and why these were rejected in favor of the proposal ultimately selected. Nor are such programs accompanied by background papers stating the important technical considerations, evaluating the benefits and costs, and identifying the risks arising from lack of technical information or understanding. When such studies are performed by Executive agencies, they are rarely made available to Congress (let alone the public). So up until now, not having the time or staff to perform parallel investigations, Congress has not been able to fulfill its "checks and balances" function. The Office of Technology Assessment, if adequately financed, will enable Congress to respond more intelligently.

Interest in technology assessment is not restricted to the federal level. There is talk among state and city officials about it, and surely some of them will acquire this kind of capability.

The Prospects

Certainly the need exists and surely technology assessment is potentially a very useful instrument. Whether it evolves as such is another matter; it is still in the formative stages, and it has its skeptics, even opponents. There are those who argue that the present system has survived the test of almost two centuries of use; why tamper with it? And, too, there is some fear that rationalization and systemization are taking over from good old-fashioned politics. But when multibillion-dollar investments are involved, the physical well-being of millions is at stake, far-reaching and oftimes irreversible social and environmental effects are inevitable, do you want lobbies, pressure politics, and other aspects of the process you don't read about in schoolbooks to have the influence they now have? Other opponents see technology assessment looming as a pinch in the pipeline of progress and, for that reason, they are prone to refer to it as "technology arrestment."

Technology assessment is no panacea. Although it will provide significantly better information than is now generally available to decision makers, surely it is not going to eliminate mistakes. Appraisals of costs, benefits, and risks are not going to be completely factual. Furthermore, there is no need to fear the elimination of human inputs—there will still be call for judgment, debate, compromise, and expert opinion.

Up to this point government has received most of the attention, which may well prompt you to conclude that private enterprise is getting off too easily. But it is not off the hook yet; the next chapter applies primarily to the private sector. Furthermore, there is sufficient reason for my emphasis on political decisions in the technical area; government has become more influential than private enterprise in many such matters. A majority of the most significant and controversial engineering ventures today—new freeways, urban mass transit, proposed jetports, weapons systems, SST's, urban renewal, space exploration. . . , are government doings.

EXERCISES

1. On the basis of current usage and the number of ships that are too large for the current Panama Canal, a new and enlarged canal in the same general area seems inevitable. This project is mainly in the talking stages at present, and most of the talk is about a

sea-level canal. The present facility has locks and a freshwater lake that offer a barrier to the flow of sea life, but the sea-level version will permit the free exchange of species between the Atlantic and Pacific Oceans. This is one of a number of significant effects of this proposal.

This venture is a natural for technology assessment. Draw up the specifications of such an assessment, indicating what *questions* you believe should be answered before a decision is made. Assume that you are instructing a team that is about to undertake an assessment of this venture. Refer to the January 29, 1971 and September 26, 1969 issues of *Science*.

2. NAWAPA (North American Water and Power Alliance) is the proposal for capturing Artic fresh waters and distributing them through Canada, the United States, and Mexico by an elaborate system of canals and aqueducts. It's intriguing, but it's also expensive, with a price tag of around $100 billion, the most ambitious construction project that man has ever seriously considered. It would provide enormous amounts of electrical power and many miles of navigable waterways in addition to the fresh water benefits. According to its proponents, the revenues would be substantial, eventually recovering the original cost.

Although it looks attractive, we have learned by now that such schemes should be investigated thoroughly before proceeding and, for that matter, before rejecting. Therefore, on the assumption that you are a member of a tri-nation technology assessment team, what matters do you say should be investigated and what questions should be answered? (Refer to *Power Engineering,* January 1967.)

3. Observe the pattern set by exercises 1 and 2, and then perform the same function for some other major engineering venture, such as:

(a) **Satellite transmission of television signals over locations like Europe or Africa.**
(b) **Man's move onto the sea, which embraces, among other things, the extraction of food, oil, minerals, and energy, construction of large structures, and placement of weapons systems.**
(c) **Adoption of a nationwide "electronic cash" system, in which the bulk of our transfers of funds (especially those now executed by check) would be handled by computers and wire transmission.**

(d) A venture specified by your instructor. Regardless of the particular undertaking, the important matter is that you give some thought to the kinds of questions that assessors should answer before man embarks on a large-scale engineering venture.

RESTRAINTS

Chapter 19

In the private sector the counterparts of public officials who make type I and type II decisions come in for their share of criticism. But for some different reasons. Unlike their political counterparts, decision makers in private enterprise cannot readily shy away from the matters of goals and priorities, nor can they be consistently guilty of tunnel vision with respect to alternatives. Neglect these matters for long and the management, if not the whole business, will go.

Their biggest failure is in cost accounting, using the term in a broad sense. Manufacturers ordinarily let society assume the cost of properly disposing of the wastes they dump into streams, lakes, and atmosphere. Who do you think strip-mining companies have assumed will bear the costs of devastated countrysides? Similarly, the cost of properly disposing or recycling of consumer goods at the end of their useful life is ordinarily charged to society. This practice of ignoring certain "external" costs for which they are never billed affects producer and consumer behavior. Many of their decisions would be affected if buyers paid true costs. The true cost of electricity would include what you are now paying, plus an assessment for all environmental damage inflicted from the coal mine to the outlet in the wall, plus any other costs now being borne by someone other than the consumer. Many people would behave differently if they simply *knew* the true costs of items like electricity or their automobile.* There is no danger of that, however; true costs for all practical purposes are unknowable.

Costing practices are not the only criticizable aspect of type I and type II decision-making procedures in the private sector. There are businesses that market some pretty lethal products, and

* Morgan, Barkovich, and Meier, "The Social Costs of Producing Electric Power from Coal: A First-Order Calculation," *Proceedings of the IEEE*, Vol. 61, No. 10, October 1973. This is an admirable and revealing attempt to quantify the costs to society not presently included in the price of electricity.

surely there are other ways in which private enterprise abuses man and nature through technology. But most of these abuses can be traced back to their particularly vulnerable costing practices.

There may be some who feel that private enterprise can eventually be persuaded to assume the costs of the public resources they consume, pay for the damage they do, and otherwise absorb costs that they have heretofore "externalized." The more realistic critics of businesses that palm off sizeable costs on society think otherwise.

The Need for Restraint

Man has acquired some remarkably powerful capabilities that often prove to be double-edged. The bulldozer has certainly made its mark as a benefactor *and* devastator, and television, which is a great source of entertainment and enlightenment, has frightening potential for influencing public opinion and social values. The device for releasing the power of the atom is perhaps the most dramatic of man's many double-edged contrivances. Although not inevitably so, these and other technical developments do, in fact, spell ill for at least some of mankind. Even contrivances born of the best of intentions sometimes are mixed blessings.

So technology while it provides man with an opportunity to better his life and lot, has increased the variety and potency of his opportunities to abuse. Several types of abuse, in fact. One is the infliction of harm directly and deliberately, as with weapons. The majority of technological abuses, however, are of man and nature as by-products of the use of technology for otherwise innocuous purposes (e.g., smoke from the factory's stack).

Commercial deep-sea fishing has become so efficient, thanks to technology, that relatively few fishermen are now able to sweep the oceans free of commercial fish. And this is what the fishermen of all nationalities seem bent on doing. Engineering has provided them with tools that serve an important and highly useful purpose, but which they can also use to destroy a valuable resource. Such exploitation is hardly considerate of future fishermen, to say nothing of future, underfed masses. What fishermen are doing to fish resources you can *see* other private interests doing to public resources like the atmosphere, waterways, and land itself. This is nothing new to you.

A major upshot of man's enlarged capacity to commit "technological sin" is the need for greater restraint in what individuals and institutions do with their machines and structures. Some of the reasons modern engineering creations are more consequential were cited in Chapter 16, (e.g., power, population and dependency). The streets are more crowded, and so are the highways, lakes, rivers, and airways. With planes, guns, automobiles, industrial effluents, and so forth, it is the increased power to inflict harm combined with greater population density that create the need for more restraint.

True, too, we are much more dependent on technology for fulfillment of needs and wants; the public is now at the mercy of a variety of technically based enterprises, power companies for example, so some government regulation of them is necessary. Indeed, this has already come to pass for some industries.

But power, population, and dependency are not the whole story. There is this matter of complexity; it continues increasing,

Figure 19-1 *Don't think this isn't going to happen again!*

conspicuously so in consumer products; the automobile, washing machine, and even the bicycle are illustrations of this. As a consequence, it is becoming more and more difficult for the buyer to appraise intelligently what he is about to purchase, to anticipate hazards in operation, and to make repairs and adjustments. Thus, the consumer is more often at the mercy of the producer.

And so technology provides more than ample opportunity to abuse; in fact, when it comes to powers to contaminate, maim, clutter, and influence, man has never had it "so good." Who disagrees that some restraint is necessary? The issues are how much and what kind. Concerning how much: there is a sound case, based on the power, population, dependency, and complexity arguments for greater restraint, and certainly there are strong political pressures in that direction (see below) but how far we should go in this respect is going to be a rough problem to resolve. This question, and questions concerning the most effective type of restraint (self versus legal versus any other restraining mechanism that shows promise), I refer to as type III decisions.

Mounting Pressure

A conspicuous manifestation of the prevailing feeling that man must exercise greater restraint in what he does with his technology and where he does it is the growing resistance that citizens are offering to certain technical ventures. Certainly those who misuse technology are hearing about it, especially from the environmentalists and consumer advocates. The resisters are becoming more numerous, more vocal, better organized, and more effective.

These voices of protest serve a restraining function. There are proposed power plants, pipelines, freeways, and other projects that are being resisted by ad hoc citizen groups. In aggregate, these opposition groups serve as a retardant, but not necessarily the best kind. Fighting virtually every proposed construction project on an ad hoc basis has been a useful interim means of restraint. However, among the drawbacks of this approach are the prolonged legal proceedings associated with these bouts between backers and opponents; they are trying and expensive to the combatants and society. This is one reason many are unhappy with this method of restraint.

But there is another reason. The ad hoc opposition to major, one-shot technological acts as a restraint mechanism is unsuccessful in some instances, inapplicable in others. Citizen opposition by itself has not successfully curbed pollution by industry, municipality, or automobile. There was little response to protest until legal restraints were enacted and offenders threatened or prosecuted. Furthermore, there are many misuses of technology that are repeated acts by individuals with their automobiles, snowmobiles, and guns, to which direct citizen opposition, practically speaking, doesn't apply.

Alternative Restraint Mechanisms

There are a number of ways of preventing something like a complete wipeout of the sea's fish resources. One is "self-restraint"—adherence to sensible conservation practices by commercial fishermen. An alternative is the "bill them" approach, under which fishermen are taxed according to their catch and these funds used to support fisheries and other means of sustaining fish populations. Another is enforcement of legal quotas, intended to conserve this resource, by an international agency. These three alternatives—voluntary restraint, billing abusers for damage done, and government-imposed restraints —present themselves in most instances where technological abuse is likely. Which of these will predominate in the long run?

You probably have an opinion, and so do many others. Here is where the weight of recorded opinion lies. Virtually everyone is pessimistic about self-restraint. Most agree that the opportunities for abuse provided by technology are too numerous and too tempting; too much so for humans, frail as we are. You cannot depend on individuals to refrain voluntarily from using these powers to do in their fellow men or do them out of something. And it is naive to expect business interests to refrain voluntarily from using technology to exploit national resources or the public. Given the opportunity, man will exploit man and man will exploit nature. Of course, that opportunity has always existed, but never were the means for exploitation what they are today.

"Billing them" is a possibility. Through taxation or a more direct means of assessment, abusers pay for damage done. An example is government assuming the responsibility for restoring land defaced by mining companies and billing the companies accordingly. This is not a fine; here the abuser may continue defacing the countryside or polluting or whatever, but *he* pays. Although this method has a certain appeal (it is a step toward true costs), it receives little support in the literature or in practice, probably because it is a cumbersome mechanism and because of the difficulty of assessing the damage (e.g., to residents near a jetport)!

So if there is so little faith in voluntary restraint, if economic disincentives are considered impractical, what prevails is the method of last resort—legislated restraints.

Legal Restraints

Governmental restraints on the use of engineering contrivances are nothing new, nor are regulatory agencies for this purpose. Speed laws have been around for some time; regulation is certainly not new to the electric power industry; there already are laws controlling the use of electronic bugging devices. The issue is one of degree, touchy to be sure, that will not be resolved overnight or without controversy.

The pressure on government to legislate against "technological sins" is mounting. The cry for tighter pollution controls has been extended to auto features, bugging devices, motorboats, guns, and a wide variety of other consumer products. Tighter curbs have also been demanded on private exploitation of scarce resources like water, the marshes, minerals, fish and fowl, and open land, most of which are consumed by engineering creations of one type or another. Tighter safety regulations in mines, factories, and construction, where machines are used extensively are also being requested. Tighter legal restraints appear to be the only practical alternative in many instances. With this prospect, it is important to say something about two key decisions legislators face. One concerns *timing,* the other *degree.*

Setting Legal Restraints—The Timing

The snowmobile provides a timely illustration of what is becoming a familiar pattern of events. While this "sport" is becoming popular, participants are relatively free to use their machines where and how they please, at least long enough to become accustomed to this freedom. But not without unfortunate consequences for man and nature, so that some restrictions on features and uses of snowmobiles seem inevitable. You can imagine, though, now these would be welcomed now.*

What is happening with the snowmobile is getting to be a familiar story. It begins with the acquisition of a new capability

* You might enjoy the article "Cold Show Out in the Cold Snow" from *Sports Illustrated,* March 16, 1970. Author Jack Olsen makes a case for tighter control of snowmobiles.

(e.g., for transporting gas through a pipeline or mechanically harvesting fish), the use of which is practically unrestricted, at least at the outset. But exploitation creeps in and continues unabated, perhaps unchallenged, until society eventually and belatedly responds by pressing for government controls to curb abuses. But this time lag between the time a capability becomes available and the imposition of restraints causes big trouble. Typically and understandably, the individuals or institution using that capability have become accustomed to freedom in doing so; a precedent has been set.

In the case of pipeline companies, mine operators, or automobile makers, however, it is more than a matter of precedent. After it has invested millions, an industry is not eager to rebuild the system according to a new set of rules. This is true of the airline industry; although their engines are noisy, it would be unreasonable to *now* expect them to invest hundreds of millions to replace them with quieter engines even if they were available.

The point is that we should be doing a more thorough job of studying new capabilities *as they evolve,* in an attempt to anticipate the eventual need for restraints, so that:

- **The cost of enforcement can be predicted and included as part of the *total* cost of a new capability, thereby influencing the decision to proceed with its development. Hopefully we won't invest in development of a technical capability if we know in advance that it is going to be miserable or impossible to live with.**
- **The necessary regulatory legislation can be enacted *before* large capital outlays are committed, before "rights and privileges" become established, *and* before serious damage is done.**

Thus, there is another important purpose to be served by technology assessment: anticipation of the need for restraints, for costing purposes and so that laws will be on the books when the new capability blossoms into widespread use.

Setting Legal Restraints—The Trade-Off

Sooner or later, like it or not, society will be forced to accept closer regulation of what private and public enterprises discharge into the environment, of materials manufacturers can use, of energy consumption, of automobile features, of. . . . But there is a limit; underrestraint can eventually turn to overrestraint. In some areas this appears to have occurred already. The motorcycle helmet law prevalent among states is generally viewed as overrestraint. And what about the seat belt interlock fiasco in 1975 automobiles? Or bring up the matter of governmental regulation with an executive of a railroad, electric utility, or airline. Obviously a trade-off is called for, a balance between ridiculous extremes, one being a laissez-faire policy that permits us to hang ourselves, and the other being an overrestrictive policy that stifles human freedom and economic initiative. Specifically, the trade-off is: the cost of physical, social, psychological, and other damage we do with our technology versus the cost of creating, enforcing, and complying with restraints *as well as* of loss of freedom. Since you know something about optimization and trade-offs, you recognize the prime objective here: to find a degree of restraint that minimizes total cost to society.

You can view this trade-off in terms of freedoms. Individually and collectively, citizens must resolve a conflict between two types of freedom: the traditional freedom that relates to belief and behavior versus freedom from abuse, exploitation, bodily harm, or even annihilation in an overpopulated world equipped with underrestrained powers. The big challenge is to develop effective means of controlling these powers without unduly encroaching on man's traditional liberties.

Legal Restraints on Engineers

Government affects engineering in numerous ways, but surely the most meaningful in day-to-day practice is the accumulation of laws and legal precedences within which engineers must operate. You are somewhat familiar with the changing liability attitudes and growing influence of the courts, these came up in an earlier discussion of product safety. But you may not appreciate to what extent government influences the practice of engineering through legislation.

One example is the Occupational Safety and Health Act (OSHA) of 1970. Basically, it established a host of occupational health and safety regulations plus an ongoing procedure for adopting additional regulations as the need is perceived. Especially if you are an engineer designing or supervising the operation of machinery for construction, transportation, manufacturing, electric utilities, agriculture, lumbering, longshoring you name it, will you experience the strong influence of this Act. For instance, for designers of cranes, it specifies safety factors for cables, shielding requirements for moving parts, requirements for steps, ladders, and hand-holds, limits on noise, and so forth.

Another illustration is the 1970 Clean Air Act, with which you are probably generally familiar. You can visualize what it means for engineers. Then there are the Consumer Product Safety Act, Noise Control Act, Interstate Commerce Act, Federal Water Pollution Control Act, and Communications Act, just to name a few.

Legislation such as this directly establishes restraints on the practice of engineering through codes, performance standards, operating regulations, limits on noise, limits on discharges into the air and water, and even through requirements for certain procedural matters concerning contracts and construction permits. But that's not all. Most such legislation also creates regulatory agencies, examples are the Occupational Safety and Health Administration, the Environmental Protection Agency, and the Consumer Product Safety Commission. These agencies generally have enforcement, interpretive, and regulation-setting functions that tend to make them more influential than the original legislation.

Thus, the federal government's heavy hand is felt by engineers through its type I, II, *and* III decisions. In type III decisions, state governments also play a major role. Considering all the legislation, court-established precedences, and regulatory agencies at federal and lower levels, government's influence over engineering is considerable. Recall, returning to a point made at the outset of Part 4, that the influence is two-way: what engineers do affects government (generally indirectly), and government affects what engineers do.

Managing Technology

Choosing what capabilities to develop (type I decisions), deciding how to apply existing capabilities (type II), and determining what we dare not do with them (type III) are the main decisions through which man *manages* his technology. The term *technology management* and what it implies are indeed appropriate. Hopefully, both will soon be widely accepted. In addition to the three decision types, it includes feedback mechanisms that enable decision makers to alter their policies, regulatory practices, and investments in research and development as the economic, social, and environmental well-being of society dictates (Figure 19-2).

What governmental structures will best perform this management function is a complicated matter; thinkers will labor over this matter for some time. However, it *is* clear that careful planning, effective feedback, a more distant time horizon than decision makers are accustomed to using, and a constituency of men everywhere are essential features of a successful technology management capability.

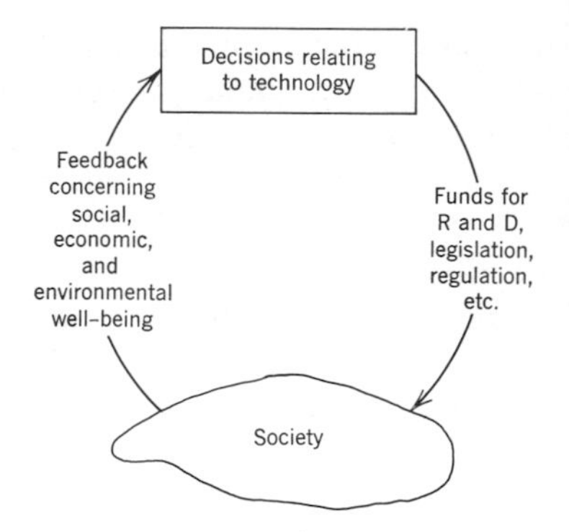

Figure 19-2

EXERCISES

1. The day of this writing the headlines are announcing another tragic midair collision, in this instance between a single-passenger private plane and a jet airliner, killing 83 persons. The traffic controller could have prevented this if the small plane had shown up on his radar screen, which would have been the case if the plane had been equipped with a small electronic device called a transponder. Moreover, these planes probably would not even have been dangerously close to each other if there were stricter regulations on where small private planes and large transports can fly. Relate this incident, similar cases, and the prospects of same to the subject of this chapter. Specifically, what is the connection, what issues are raised, and what trade-off must be made?

2. Do you foresee legal restraints on uses of computers? Why, and on what kinds of applications? What are some of the counterarguments? (You may find it helpful to do some extra reading. *Technology Review* has carried some thought-provoking articles on the subject; so have computer-oriented periodicals.)

3. One of the more curious and controversial restraints on man's use of his technology is the popular requirement among states that motorcyclists wear helmets. Can you explain it? Can you justify it? What issues does this requirement raise?

4. The federal government has an increasing say concerning automobile features, such as emission levels and safety features. The trend in these respects is obvious. In safety, first it was mandatory safety belts, then it was "reminder" devices that discourage you from driving without a fastened seatbelt, and then devices that prevent you from starting your car before your belt is fastened. These restraints imposed by the federal government are controversial, to say the least. What is your reaction? What is your solution? Comment on the issues raised.

5. Compose a letter on behalf of affluent Americans, addressed to the economically and materially deprived populations of Asia, Africa, South America, and other areas of the world, explaining to them why they can never have it is as good as we do materially. And don't kid yourself; these people are aware of our affluence and they want in. Yet, to raise their standard of living they must increase their energy consumption and level of industrialization, which will dump more junk into the atmosphere to add to what we Americans are already contributing. As it is, we, who constitute only 6 percent of the world's population, are now contributing approximately half of the contaminates of the world's atmosphere. By simple arithmetic, you can extrapolate what the world's population would do to the atmosphere if it reached our level of material development. Something's got to give.

Of course, if you're going to write any letter on this subject, it should be addressed to affluent Americans explaining why *they* must eventually cut back on their abuse of the atmosphere and on their wasteful consumption of energy and material resources, so that millions of others can improve their bodily, economic, and material lot, a worldwide leveling, so to speak.

So, compose a letter to affluent Americans, making a case for an "adjustment" in their standard of living and level of resource consumption, in the interest of avoiding future environmental catastrophe, social chaos, or both.

Attitudes, Biases, and Remedies

PART 5

Opinions of the works and workings of engineers are indeed diverse. Surely you have noticed that such views range between wide extremes and that they are almost invariably based on a limited perspective. But you may not be conscious of the prevalence of ambivalence in our attitudes toward technology, its developers, and its appliers. Part 5 explores these attitudes, their origins and consequences, and means of tempering at least the extremes.

BOON OR BANE?

Chapter 20

Over the recent past many words have been written and spoken, notably by the well-known British author C. P. Snow, about a gap between what are rather loosely described as the literary and scientific cultures. Some feel that Snow's dichotomy is an oversimplification; perhaps they take him too literally. However, by precipitating much thought, many publications, and vigorous debate over his "two-culture gap," Snow has performed a service by focusing attention on the diverging values, communication failure, and other factors that separate the technical and nontechnical worlds.*

Although this gap between scientific and literary intellectuals disturbs some, of greater consequence is a related gap, on the one side of which are Snow's literary intellectuals along with *all* people who harbor a dim if not disdainful view of the whole of modern technology and its implications, the *technophobes*. Opposite them are the *technophiles*—advocates of technology, consisting of a majority of engineers and physical scientists, and it seems, a sizeable proportion of the public at large.† (I am taking the same risk C. P. Snow took: that readers will take this simple dichotomy too literally. For one thing, people tend to be ambivalent in their valuation of technology. And another reason that you cannot readily type everyone as technophobe or technophile is that many people, maybe a majority, have no particular feelings about the matter.)

These two "cultures" are distinguished in part by a difference in attitudes, described as a conflict of biases. It is a conflict worth exploring in detail.

* C.P. Snow, *The Two Cultures: and A Second Look,* Cambridge University Press, 1965.

† *Science Indicators,* National Science Foundation, 1973, reporting on an independent survey of 2200 adults representing a cross section of the population.

Attitudes: Technology in General

Mainly among intellectuals, there is a running debate over technology's net worth and future prospects for mankind. Sample the writings on these issues and you will be confronted by a broad spectrum of viewpoints with these extremes.

Voice of the Ardent Technophile

In the past we have been showered by rosy reports of accomplishments, glowing predictions of material achievements to come. Surely you know the type. They speak of interplanetary space travel, a computer in every home, videophone communications, high-speed transportation in individualized capsules, the all-electronic kitchen—a utopia of physical conveniences and comforts that brings to mind a technological horn-of-plenty spewing material fruits. Persons with views at this extreme suffer from acute technophilia, the symptoms of which are overly optimistic expectations of the benefits attributable to technology, insensitivity to adverse effects, and a materialistic bias.

Highly optimistic views of the contributions and potentials of technology have been expounded for a long time, reaching their peak probably in the 1940s and 1950s. But such utterances are no longer so numerous or so bold. As the voice of the technophile fades, the voice of the technophobe grows louder.

The Counterview

Over the last decade technophobes have become busier with pen and podium. Should you sample their writings, try the works of Jacques Ellul, noted French philosopher-historian, idol of intellectual technophobes. He writes with what technophiles certainly consider a pessimistic pen. Ellul, incidentally, holds an unusually broad view of technology. In his words it is "the totality of methods rationally arrived at," and so to him it obviously means more than physical contrivances. In the English translation of Ellul's French version, his very broad notion of technology is referred to as *technique*. He sees technique as a serious threat, as a force unto itself, as no longer controllable by man. He speaks of this autonomy of technique in this way.

> The important thing is that man, practically speaking, no longer possesses any means of bringing action to bear upon technique. He is unable to limit it or even to orient it. I am well acquainted with the

> claims of those who think that society has technique under firm control because man is always inventing it anew. I know too of the hopes of those who are always prescribing remedies for this sorcerer's apprentice whom they feel free to invoke without discernment. But these claims and hopes are mere words. The reality is that man no longer has any means with which to subjugate technique, which is not an intellectual, or even, as some would have it, a spiritual phenomenon. It is above all a sociological phenomenon; and in order to cure or change it, one would have to oppose to it checks and barriers of a sociological character. By such means alone man might possibly bring action to bear upon it. But everything of a sociological character has had its character changed by technique. There is, therefore, nothing of a sociological character available to restrain technique, because everything in society is its servant. Technique is essentially independent of the human being, who finds himself naked and disarmed before it.*

Ellul has company; there is Mumford *(The Myth of the Machine)* and Marcuse *(One-Dimensional Man),* but no one seems to excel Ellul, referred to by critics as gloomy Ellul. A pattern is evident in such writings; they speak of dehumanization of man at the hands of machines, of stimulation and gratification of a materialistic compulsion, of man's bewilderment in a technical world, of the subversion and suppression of traditional values, and of technology becoming an end in itself. They paint a bleak picture of mankind's present plight and future prospects, contrasting them with illusory visions of a simpler, happier life in the preindustrial past. Jacques Ellul, his kind, and his disciples are philosopher-technophobes, on the extreme fringe of the antitechnology movement, to be sure. But there are other technophobes of varying shades and differing orientations.

Consequences

The effects of bias are familiar; among other things, it inhibits rational response, renders debate futile, and impedes communication. If you suffer a pronounced bias of either type described, you most likely cannot respond objectively and intelligently to issues, proposals, and policies involving technology—in your job or in your citizen role. This is one of the reasons I hope you can avoid contracting an acute case of technophilia or technophobia.

You also know from your own experience that biases inhibit

* *The Technological Society,* Jacques Ellul, Alfred Knopf, Inc. 1964.

communication. Nowhere are the biases concerning technology more conspicuous and consequential than on campuses, where you will have no difficulty observing the communication gap between technically oriented and liberal arts faculties. To make matters worse, these attitudinal differences and the communication gap are mutually aggravating. Opposing biases inhibit dialogue. This, in turn, does nothing to improve understanding, which aggravates the attitudinal difference, which reduces the chances of dialogue, which. . . . It all adds up to a hard-to-break cycle.

Of course, there are communication barriers between all intellectual disciplines, so why make a special to-do about the one between technical and nontechnical segments of the campus community? Because, although their numbers are small, they mold the biases of many, and they continue to do so in an age in which convergence is more desirable than ever.

A Reasonable View

When an author describes what he considers a reasonable view, he reveals his own position. (You should wonder about that since, only by having a ''bias reading'' on an author, can you intelligently interpret his commentary. That holds for all commentary.)

I have always respected the thinking of Bertrand de Jouvenel, noted French philosopher, who offers these thoughts on what technology offers to man.

> I deliberately refuse to think of technological progress as the cause of change in men's ways of life. I deliberately choose, rather, to think of it as the chief means of such change. The idea that mankind is being and will increasingly be driven along a determined path by the gathering forces of technology seems to me a nefarious idea, generating in the minority an impotent and paradoxical technophobia, and in the majority the same kind of blind ecstatic confidence in what shall come to pass that was characteristic of laissez-faire in the narrower field of economics.
>
> Being ''against'' the increase of potentialities afforded by technological progress seems silly. The only conceivable excuse for such an attitude lies in the contrary and complementary silliness of those who turn technology into an idol, bountiful but imperative, to be served unquestioningly, instead of seeing it as the slave evoked by Aladdin's

> lamp, available to serve us to our profit or to our undoing, according to the wisdom of our choices.*

True technology puts unprecedented powers at man's disposal; indeed, some of these have been used for exploitation; surely good judgment is not always shown in application; obviously there have been unfortunate side effects. But all in all, man has done well by technology. To be unaware of the benefits is to be unobserving; to refuse to acknowledge them is to be downright ungrateful.

Concerning the future, many people reject the extreme pessimism of the Ellul-type technophobes, especially since there are signs that man is growing wiser in the handling of his technology. It seems that the romance with science and technology is tempering, with the demise of the SST venture as the apparent high-water mark. Finally, gradually and significantly, we are becoming more sensitive and sensible as we make our type I, II, and III decisions. Surely there will be no moratorium on technical advance; it is not necessary; it is not desirable. Paul Ehrlich and John Holdren writing in *Saturday Review,* argue that mankind will require more, not less, science and technology, but certainly not always the same kinds or on the same scales we are accustomed to.

> It is beyond dispute that technology is an essential ingredient if the bulk of mankind is to be provided a decent existence. But this does not mean that the answer is simply more of the same kinds of technology that have helped to create today's predicament, and it does not mean that a particular technological scheme is wise simply because it is well-intentioned. We need more transportation, but fewer automobiles. We need more housing, but less suburban sprawl. The world may need more aluminum and steel for communications networks, bridges, and railroads, but the United States certainly does not need more beer cans.†

Man can, he should, and he will continue advancing technologically and, in the view of many, with greater discrimination.

* From "Some Musings," by Bertrand de Jouvenel, a chapter in *Technology and Human Values,* Wilkinson, Sykes, et al, published by the Center for the Study of Democratic Institutions, 1966.

† Paul Ehrlich and John Holdren, "Technology for the Poor," *Saturday Review,* July 3, 1971.

Not everyone is optimistic. "Can technology be more humane?" is a question frequently belabored on paper and podium.* The response of the Ellulian school is no. But why debate? The question is inappropriate. It implies that man's tools, know-how, and organizations are separate from and opposed to man himself. Nonsense! Are *humans* not running corporations and government? What harm is an automobile until a *person* gets behind the wheel? Electrical appliances are purchased and operated by *people*. People are in charge—as decision makers, purchasers, operators, . . .

The appropriate question is, "Will *humans* be more humane?" As Melvin Kranzberg puts it: "The question, therefore, is not whether man can master technology, the question is whether man can master himself," a message that also comes across in these words by Peter Drucker.

> It is a poor carpenter who blames his tools says an old proverb. It was naive of the 19th-century optimist to expect paradise from tools and it is equally naive of the 20th-century pessimists to make the new tools the scapegoat for such old shortcomings as man's blindness, cruelty, immaturity, greed, and sinful pride.
>
> It is also true that "better tools" demand a better, more highly skilled, and more careful "carpenter." As its ultimate impact on man and his society, 20th-century technology, by its very mastery of nature, may thus have brought man face to face again with his oldest and greatest challenge: himself.†

By more humane I mean exercising greater restraint in what we do with smokestacks, autos, television, wastes, and airplanes; showing greater compassion for the victims of automation and urban freeways; and showing deeper concern for the needs, wants, and priorities of people.

The Engineer's Image

What about attitudes toward and knowledge of the prime conceivers and developers of technology: engineers? There are some rather extreme attitudes and common misconceptions at work

* For instance, Paul Goodman, "Can Technology be More Humane?" *New York Review of Books,* November 20, 1969.

† Peter F. Drucker, writing in *Technology in Western Civilization,* edited by Kranzberg and Pursell, Oxford University Press, 1967.

here, too, concerning the qualities, work, role, and influence of engineers. Of course, much of this book is devoted to enlarging your understanding of these matters but, up to now, little has been said about impressions of the engineer's power and influence in the affairs of men. There are some prevalent misconceptions of him and his role.

It is true that engineers have bungled jobs; no, we don't speak out enough on public issues; yes, we do sometimes fail to take a stand against our employers' practices when balking is called for; no, not enough of us participate in public affairs; yes, we sometimes fail to give due consideration to the indirect effects of our creations; yes, we should take broader views of problems we solve; of course we are not perfect. On the other hand, we can't be to blame for *all* technological sins and blunders, as some critics will have you believe. Supposedly, we are responsible for pollution of the environment, deficiencies in public transportation, highway blight, congestion, and a host of other environmental and social ills. Ridiculous, and here's why.

Let's face it, engineers do have *some* influence in these matters, but we are not in charge. The mayors, governors, legislators, and agency heads are still on the job. Corporate managements and boards of directors are still running their shows. Furthermore, there is a glaring inconsistency in the criticisms of engineers. One group, by overdoing it with the blame, implies more responsibility than we have. Another group views us as blameless flunkies, implementors of the decisions of others. To the critics who through blame flatter us with implied responsibility, we say *pay us for it*. And those who see us as lily-white need to be reminded that decisions made by executives and public officials are often based on proposals, advice, and information supplied by engineers. Furthermore, we have an obligation to balk when our employers' decisions are contrary to the public interest.

Of course, it is ridiculous to blame anyone. But if you must, remember that politicians, architects, vested interests, corporate executives, and an apathetic public are eligible, too. But anyone who fingers *a* scapegoat for situations like our deteriorating environment is naive (or thinks the rest of us are.) To say that engineers or politicians or industrialists are responsible is a gross oversimplification. The forces at work as decisions are made in the public sector are much too complicated for blame-putting to be that simple.

When it comes to engineering decisions in private enterprise, responsibility is almost as difficult to pinpoint. If you are under the impression that engineers make a corporation's engineering decisions and are therefore responsible for the company's technical sins, you, too, are crediting us with power and influence we wouldn't mind having! Most major engineering decisions must pass through a hierarchy of corporate decision makers, and although we wish at times it were otherwise, engineers do not make company policy or control corporate purse strings!

And on the subject of blunders, perhaps you were thinking that we are different in this respect from the other professions. In one sense we are: engineers' failures frequently become full-view monuments, like the ill-located freeway, unsightly water tower, or aborted space shot. Here physicians, lawyers, teachers, and psychiatrists have it made. Their blunders go unrecognized or under the rug, rarely on public display. No profession makes claims to perfection.

Frankly, if you know the score and are relatively unbiased, you can only conclude that on the whole, engineers are not the roots of all technological evil, nor are they the noble models you read about in high school career brochures.

EXERCISES

1. Not everyone agrees with the position of Kranzberg, Drucker, and others that "the enemy" is man himself, not his technology. Ellul is one. You might be another. What is your position?

2. There is a school of thought that maintains that modern technology leads to an objectionable, to them repulsive, sameness of human attire, habitat, work, and the like. Is this generalization warranted? On what basis do you agree or disagree?

3. Here is an item from *Time* magazine entitled "No Way Out, No Way Back," which appeared shortly after a major snowstorm that pretty well paralyzed things in the New York metropolitan area.

> What will it be like, at the finish? In *Weekend,* French Film Maker Jean-Luc Godard foresees the end of the world as an immense traffic jam. Stanley Kubrick sees the men of *2001* as murder victims of a machine they have made more clever than themselves.

Or consider this scenario: The people are thrown together against their wills, trapped in colossal, modernistic buildings on a landscape devoid of trees. The lights are always lit. Pavement stretches everywhere. Cars and buses and trains and aircraft are useless: there is no way out. No darkness. No silence. No beds. No escape from an endless series of broadcast announcements, no avoiding the silly, circular games of other people's children. There are queues for food, queues for asking questions, queues for liquor—and finally queues for nothing, because there is nothing left. Then there is only boredom, and the debris of boredom. Dirty glasses, old newspapers, crumpled cigarette packs. Even the people are debris. Women wander aimlessly, their hair frazzled, their makeup so streaked that their faces look as if they are melting. Men in rumpled suits, with three days' growth of beard, slump in chairs staring at the message boards that bear no messages.

Packaged and Shipped. Perhaps it will all begin with a simple and foreseeable act of God—say a heavy snowstorm in New York City. There, last week, at the world's largest international airport, the scenario came true. Even at its best, an airport terminal seems inhuman—a monstrous machine disguised as a building and designed to process people and baggage. To the machine, there is no difference between men, women, children, suitcases, pets. All are collected, screened according to route, classified by status, divided into units of the right size, packaged in aircraft—and shipped. When 17 inches of drifting snow clogged the runways and access roads of John F. Kennedy airport, 6,000 people were forced to exist inside nine broken machines. And, because of the incredible slowness of Mayor John Lindsay's snow-removal machinery, they were prisoners there for three days.

For Michael Rogers, a student headed back to Georgia's Oglethorpe College, the ordeal began shortly after 10 A.M. Sunday, when he telephoned Eastern Airlines to check on its 11:25 A.M. flight to Atlanta. Assured that the flight would depart with "a slight delay despite the snow," Michael drove to the airport and checked into the Eastern terminal at 11 A.M.—only to discover that the flight had been canceled. He was still there 56 hours later. Thousands of other travelers were similarly misled by the airlines, which, out of either optimism or greed, led them to believe that planes were still taking off. American Airlines waited until 2 P.M. on Sunday to announce the indefinite cancellation of all future flights, although all outgoing planes had officially been grounded since 10 A.M. Eastern waited until 9:30 P.M. Sunday to announce that no flight would leave until at least noon of the next day.

Crash Pad. Passengers kept pouring into all the major terminals, only to find that the snow had left no way out and no way back. Three people never even made it to a terminal: they were found in their car in Parking Lot No. 4, dead of carbon-monoxide poisoning. As the snow kept falling and drifting, it gradually dawned on everyone in the terminals that they were completely stranded. Airline officials struggled to provide minimal creature comforts. That is, some struggled. Trans World Airlines turned out 11,500 meals and 18,500 snacks in two days. TWA's clamshell terminal building, designed by the late Eero Saarinen, proved more adequate than most as crash pad; the decorative red carpets in its gateway tunnels made comfortable mattresses for weary refugees. The airline also converted one of its planes into a movie theater, showing 3 films continuously from 10 A.M. to midnight Monday to 142 passengers at a time.

At the Pan American building, where there are no carpets, passengers stretched out wherever they could—behind ticket counters, on luggage carts, even on the huge steel turntables in the baggage area. "Everybody is taking advantage of us," complained Frank Russomanno, a salesman from San Francisco. "The cafeteria is overcharging. The airline is not considering the people—especially the children. There are 1,000 children here, and they haven't done anything for them. They should have organized games. Or something."

Eastern Airlines had only 500 blankets for 1,500 people; when a father managed to get hold of four—one for each of his children—an Eastern official demanded them back for his agents' use (the father refused). A few passengers found their way to the employees' cafeteria in the basement and stole food. As they crossed the terminal with loaded trays, they became an increasing source of frustration to 500 others who stood in line for five hours one night, only to be finally turned away. The restaurant manager blamed the foul-up on passengers, who refused to give up their seats inside, even when they had finished eating. "Some stayed for three days," he said. "They did their laundry and hung it on chairs. They refused to go." Go where?

More Chaos. A few passengers did have enough pull—or gall—to escape. One with pull was Chris Craft Chairman Herbert J. Siegel, who was stranded at the Eastern Terminal while awaiting a flight to Acapulco. Siegel called Manhattan Publicist Tex McCrary, who in turn phoned Pro Football Commissioner Pete Rozelle. They managed to commandeer a helicopter that was originally chartered to CBS-TV news. It took McCrary half an hour to locate Siegel after the helicopter landed on the Eastern runway; by the time they got back to

> the 'copter, three strangers—with gall—were placidly settled in its seats. They refused to get off, so the pilot had to fly them to Manhattan and return for Siegel three hours later.
>
> Not until almost 10 A.M. Tuesday did planes again fly out of Kennedy. By then, though, an airport access road had been plowed—creating even more chaos: in poured a stream of new travelers with reservations on Tuesday flights, who demanded that their tickets be honored. Airline agents explained that they would have to wait until stranded passengers had been cleared out—perhaps another 24 hours. Whereupon they clumped angrily out of the terminals, hailed cabs to return home, and encountered yet one more annoyance. Never noted for their resistance to temptation, taxi drivers were flagrantly gouging passengers, carrying six to a cab and charging $20 a head for the ride into Manhattan—a total of $120, or about $113 over the legal metered fare. Quite a price to pay to get from one Godard traffic jam to another.*

There is food for thought in this piece. Do you detect a bias in these words? Do you believe a majority of people are willing to pay more in order to have a terminal with a homier atmosphere? Does it make sense to stock the terminal with provisions for a once-every-thirty-years event? There is a fundamental trade-off involved here, one that humans make many times individually and collectively as they make decisions concerning nuclear plants, national defense, earthquake insurance, bomb shelters, and flood protection, for example. Discuss this trade-off. What is your overall appraisal of this article?

* *Time,* February 21, 1969.

THE ROLE OF EDUCATION

Chapter 21

It seems that when students graduate from high school, they already have a humanistic or technical bent. Surely the conditioning process begins in early youth, under the influence of family, media, and schooling. But whatever role precollege schooling may play, it is probably a college education or lack of it that does most to mold one's views with respect to technology.

When students enter college, they tend to follow their bent in selection of a major; this is only natural. Then the system goes to work, reinforcing instead of abating. Matriculants with a mild humanistic tendency, after four years of liberal studies, tend to be genuine technophobes. Similarly, after four years of technical education, students who enter with a casual techno-scientific orientation seem to become genuine technophiles. The explanation is not hard to find.

Many teachers outside of science and engineering harbor an anti-technology bias that proves contagious. Technophobic educators perpetuate disdain and false impressions with considerable leverage, contaminating generations of students with technophobia. The situation is aggravated by the fact that disturbingly few of these students are exposed to a single technical course in four years of "liberal" education. Students of the humanities and social sciences are rare sights indeed in engineering courses.

These observations on the relevance of modern liberal education come from the pen of a noted liberal arts scholar.

> Those of us who are charged with the liberal higher learning in America are unprepared to lead our students toward the historical, social, aesthetic, and philosophic understanding of the science and technology of the times. The result, in my view, is that our liberal curricula, with a few notable exceptions, have lost contact with

reality, and we are in danger of losing communication with out students. Our faculties themselves do not understand the scientific and technological forces which have produced our society. In too many cases, they do not make a reasonable effort to understand them. We are trying to make our students drink from a stagnant pond of learning, and we should not be surprised that our students find it provincial, reactionary, dated, brackish, and unpalatable.

We continue to interpret Western civilization as if it sprang solely from the Judaeo-Christian-Greek tradition, a tradition we misinterpret to exclude scientific and technological imperatives in living, thinking, earning, believing, loving, and other pursuits of ordinary life. Scientists and engineers have remade our world in the last few centuries, yet we continue to teach our liberal arts subjects with little, if any, reference to their reconstructions.

Many professing humanists mention science and technology only to denigrate them as subversive of human values. I find this particularly annoying because during much of man's history the term "humanistic" was used to distinguish what was human from what was brutal and animal. Today we find that humanism is contrasted with "science and technology" as if both were brutish and opposed to man. Yet the historical fact, confirmed by the anthropologists, is that man's attempts to understand the universe and to control physical nature through tools and implements enabled him to rise from brutish to civilized levels. Science and technology resulted from prime human characteristics and produced humane influences and changes, but they are treated as anything but that by the gentlemen of our literary, artistic, and philosophic subcultures. Only a few historians and critics, and rare philosophers of science, have sought to cope meaningfully with the actual impacts of science and technology.

I submit that our present curricula in the higher liberal learning are misinforming students about the past, exacerbating their ignorance of the present, and denying them the liberal understanding so fundamental for dealing with the future*

The awakening is long overdue. The world into which students emerge is largely shaped by (and dependent on) modern technology. Recognizing this, it seems unthinkable that our educational system says almost nothing to nontechnical students about the nature, methods, and impact of engineering. How can what is

* Melvin Kranzberg, "The Bridge—Linkages among the Humanities, Social Sciences, Engineering, and Science," paper delivered at Drexel University, November 29, 1971.

called a liberal education be liberal when it virtually ignores this pervasive force in the affairs of men?

Now the shoe is on the other foot; there is an equivalent flaw in technical education. Engineering educators seldom weave social issues, historical context, and value questions into texts and lectures. Similarly, the views of humanists and social commentators are seldom heard in engineering classrooms. This is why a colleague from "the other side of the campus," who feels that engineers are overexposed to specialized technical matters and underexposed to broadening influences, snapped, "Chase them out of their engineering classrooms and laboratories to the other side of the campus for some culture." Where has he been? This is what engineering educators have been doing for some time. At least one-fifth of an accredited engineering undergraduate curriculum must consist of courses in the humanities and social sciences. Engineering educators have long since recognized that their courses do not provide sufficient exposure to other than technical information, and they have made a serious attempt to compensate for it through this one-fifth policy. Yet not many are content with this as a remedy; engineering students' experiences in liberal arts classrooms are of limited value. Why? One reason, cited above, is that humanistic and social perspectives so rarely crop up in their engineering classrooms. Furthermore, engineering students tend to take their liberal arts courses lightly because they have never heard the strong case on behalf of such courses.

Surely these are not the only shortcomings of liberal and technical educations. Heaven forbid. The intent is only to identify educational practices that account for the respective forms of unpreparedness of liberally and technically educated people for a technological age. It is small wonder that earlier-acquired prejudices are perpetuated if not amplified in liberal arts and engineering classrooms.

Beyond the Classroom

Our system of formal education has not caught up with the technical age in which we live. But all is not necessarily lost; education can be gotten from more than schoolbooks and classrooms. In a case like this the communication media play a significant role. Over the long run, the public at large could become reasonably well informed about the goals, methods, accomplishments, failings, and issues associated with modern technology, primarily through the efforts of the press. How effectively do journalists meet this challenge?

They could do worse, and yet many of them have been guilty of distortion, careless use of terms, preying on the public's appetite for the sensational, and lack of discrimination in what they repeat. An excellent example involves automation. In the late 1940s and early 1950s, it was popular sport to publish articles that gloomily portrayed impending mass unemployment caused by the oncoming wave of automation. There was much talk about workerless factories; writers seemed to be getting their facts from one another's articles; dire predictions abounded. Some choice examples of inferior journalism similar to those described below got wide circulation.

> The popular press, inspired, in many cases by a widely-read author's prediction of a depression which would 'make the 1930's look like a pleasant joke' continued to publish extreme claims without any critical analysis, and so convinced the public that automation was a real threat. The prestigious SATURDAY REVIEW devoted its January 22, 1955, issue to ''The American Factory and Automation.'' Under the subtitle ''A Description of the New Manless Manufacture'' it quoted a senior manufacturing executive as follows: ''Our own company operates perhaps the only fully automatic factory now producing. It is a government-owned, 155 mm. shell plant in Rockford, Illinois. The plant was designed for very high production, using 20% less manpower, both direct and indirect, and . . .'' The editors blithely ignored the fact that if it used only 20 percent less manpower it certainly was not ''fully automatic'' and was hardly ''manless manufacture.'' Popular literature contained hundreds of similar inconsistencies and exaggerations. BUSINESS WEEK described Ford's new Cleveland engine plant using this title: ''Automation: A Factory Runs Itself.'' A careful study of the Ford plant in 1954 showed that over 4500 people worked there, more than 2700 of these being production workers. Surely a plant employing 4500 people does not ''run itself,'' but that was a frequent claim, often heard even in presumably carefully edited business periodicals. *

Public fears, fanned by a barrage of such literature, finally prompted a Congressional investigation that tried with little success to dispel the anxiety. Samplings of public concern indicated that the public feared automation almost as much as nuclear attack. Then, in 1966, a special presidential commission reported

* James R. Bright, ''The Development of Automation,'' appearing in *Technology in Western Civilization,* Kranzberg and Purcell, editors, Oxford University Press, 1967.

after a lengthy investigation that automation-created mass unemployment was neither upon us nor in sight, yet fears persist.

You can hardly say the press has done the public a service in its treatment of automation. In the torrent of words that were and still are printed on the subject, it is difficult for the expert and impossible for the layman to weed out the bias, faulty reasoning, misrepresentation, and just plain sloppy journalism.

And don't assume that episodes like that are a thing of the deep, dark past. We have just experienced the same thing with the environment issue. There is no doubt, we *do* have an environment problem, but environmental doomsday is hardly around the corner as we could easily expect on the basis of what we read. The editor of *Science* summed it up nicely when he wrote:

> During 1970 public concern about pollution reached an emotional peak. Many people became convinced that the environment was deteriorating rapidly and that all of us were about to choke to death from pollution. Politicians of the two major parties scrambled to establish positions on the antipollution bandwagon. Federal legislative and administrative actions that were taken will eventually result in substantial improvement in our air and waters. Convinced that the public demands cleaner air and cleaner water, American industry will spend billions of dollars on antipollution measures.
>
> An emotional peak, such as that witnessed in 1970, cannot be sustained. Earth Day activities this year were a pale shadow of those in 1969. The mass media are beginning to diminish their coverage of environmental matters, and debunking stories are starting to appear. More important for the long haul is growing recognition that environmental improvement is going to cost a lot of money and that the costs are going to be paid by everyone.
>
> The emotional peak of 1970 was built in part on a solid base but it was also built in part on erroneous information and bad judgment. We must achieve and maintain a livable environment, but we are not about to choke to death from pollution, and the world is not going to run out of oxygen.
>
> One of the odd features of the emotional peak was that it occurred at a time when most of the important components of pollution had leveled off or declined. For example, suspended particulate matter over some large cities had already decreased and carbon monoxide and sulfur dioxide content had diminished in others.
>
> Contributing heavily to the timing and the shape of the emotional

> peak was the behavior of the mass media. Reporters selectively quoted people who gave them the scary kind of story that their editors would print, or that radio and TV would use. Public emotion quickly rose. However, after a time the public interest began to level off, and the mass media are now turning elsewhere. Typically, a period of inattention will be followed by another phase in which low-key, sober assessments will provide a more realistic picture to the public.*

In fairness to journalists it should be noted that their major sources of information on such matters—the technical community and technically oriented government agencies—are not supplying the kind of information they could.

How About Government?

What do government agencies do to educate the citizenry in technical matters? Agencies like NASA *do* disseminate "technical" information—tons of it daily—through press releases, articles, booklets, films, and speakers. Regardless of the form, however, you are not surprised to learn that it is biased. These reports of achievements are part of a public relations effort intended to build a favorable public image.

So ample information is available from the government, but this "education" that the public is getting from government agencies and their industrial allies is not what the citizenry needs for its general technical enlightenment, and it is certainly not what the electorate should have to evaluate intelligently the purposes and performance of a specific agency. We are entitled to something better—for both purposes.

Where are the Scientists and Engineers?

They're around, but not very communicative, except among themselves. Scientists and engineers rarely go out of their way to communicate to the public, through the press or otherwise. When they do they are seldom considerate of their audience in their manner of writing and speaking.

Every specialist naturally uses jargon to facilitate communication with persons within his field. (I am reminded of Bob Newhart's delightful parody of conversation among space en-

* *Science,* May 7, 1971.

gineers in which he has them involved in a discussion of the *ingress-egress transfer device,* which it turns out, is the ladder astronauts used to climb between the lunar module and the moon's surface.) But that same jargon becomes an effective communication barrier to dialogue with persons outside that specialty. It's a fact of life; speakers and writers seldom translate specialized terms when attempting to communicate with laymen. This is especially unfortunate because, in engineering at least, many terms are understandable to the outsider; failure to define, not complexity, is the culprit.

Furthermore, among scientific and technical writers, it is stylish to be stiff and above all, abstruse, which does nothing to facilitate communication. Techno-scientific people tend to look down their noses at plain, forthright prose.

Of course, writing for public consumption is not popular among scientists and engineers. You do not gain professional prestige by writing books and articles for Everyman. At the universities where publishing is encouraged, if you are writing to gain "promotion points," you write for esoteric journals.

These are some of the reasons the techno-scientific community has contributed so little to the public's understanding of the workings, potential, and impact of science and engineering. Technical people argue that modern technical matters are inexplainable to laymen, partly if not mainly because they have become so mathematical. But the essence of mathematical expressions can often be captured in words, certainly sufficiently enough for this purpose. More important, laymen don't need (or care about) the theoretical intricacies of science and engineering, which is mainly where you will run into mathematical gyrations. How many people design or operate a nuclear power plant, both of which require great technical detail? What most people can know and use are explanations of general principles and terms, objective information on drawbacks and risks, identification of the basic issues and trade-offs, and the like.

It is unfortunate that our educational system, the media, and the technical community are doing so little to increase the public's technical literacy. Many view the current low level of this form of literacy as a serious national problem, including Admiral Rickover (page 278) and an Assistant Secretary of Commerce who says:

> We must raise man's understanding of science and technology. All three of our great communities—government, education and

> industry—have fallen into the error of assuming that the layman cannot understand science; that science-based technology is beyond his ken. This error has produced a widening gulf between the scientist, the engineer and the businessman on the one side, and layman on the other. It is a dangerous gulf. It must be bridged.
>
> If our democracy and our free enterprise system are to survive, the electorate must be able to understand the fundamentals of science- and technology-oriented problems which society faces; and legislators must be able to comprehend in general terms the thinking of researchers, makers and doers in science- and technology-based fields.*

What Will Help?

What can be done so that liberal arts students graduate into a highly technical world with at least a speaking acquaintance with technology? Surely sending them to the other side of the campus for some engineering courses is not a complete remedy, but it is a good beginning. Yet humanist educators must also display different attitudes in their courses. If typical liberal arts courses continue to breed disdain for the technical, if liberal arts professors continue to display an aloofness—if not contempt—for matters associated with modern science and technology, taking a few engineering courses will be of very limited value.

Of course, the "one-fifth policy" for engineering students is of limited value for a parallel reason. As social and political complexities, second-order effects, esthetics, human values, views of the technophobes, and other matters that help engineers to weave their solutions successfully into the social fabric become significant and integral parts of engineering courses, students' experiences in liberal arts classrooms will become more meaningful. Another aid is for you and other engineering students to hear from engineers, the arguments on behalf of breadth extending into the humanities and social sciences. There *is* a strong case, woven through this book, which I hope encourages you to derive maximum benefit from the liberal arts courses that you are required to take and to further expose yourself to broadening influences, whenever the opportunity arises.

* Address by Assistant Secretary of Commerce for Science and Technology John F. Kincaid, delivered to the Scientific Apparatus Makers Association, NBS Laboratories, Gaithersburg, Maryland, September 17, 1968.

So much for educational reform—except to add that there is a need for more engineers with a humanistic bent and for more liberal arts graduates to whom technical matters are not alien.

What about the "technical information" citizens receive from government? Should we not expect—demand—that local, state, and federal agencies be more conscientious and candid in keeping their constituencies informed about what they are doing, intend to do, and could be doing with technology? Obviously the public cannot be told all of the possibilities and considerations, and obviously not all citizens care or even can read, but more communication between officials and citizenry is both desirable and feasible.

Also, there is some hope of making a dent in mass technical illiteracy, *if* we can prod the technical community into communicating more to the public. Although there is mention in journals of the *need* for more engineers to participate in this kind of activity, before any amount of effective communication takes place, more techno-scientific people will have to bend a little and write for popular consumption, without jargon and rhetorical ornamentation, recognizing that the need is for more general information. This is important; society needs translators of technical matters. By becoming more fluent communicators, engineers can, among other things, help society to appraise, assimilate, and wisely use the torrent of new technology.

What Should the Public Be Hearing from Engineers?

Engineers are too tight-lipped about what they can do for mankind. Engineering has something to offer in solving a variety of pressing problems; why keep it a secret? There are very few public problems to which we have the whole answer; on the other hand, there are very few to which we have no contribution to make. Politicians and the public should know what is attainable. Furthermore, better than anyone else, we could and should tell society what the price tags are on these alternatives. How can a society set goals and allocate resources intelligently when it knows so little about what it *could* ask for, and so little about the feasibilities, costs, risks, and hazards of the things it *does* ask for? In the absence of adequate information on engineering's capabilities and limitations, goal setters can easily overlook readily attainable accomplishments and commit us to unrealistic, overly expensive goals. This is not to suggest that we can eliminate such blunders by being more communicative. But certainly our virtual silence is no help!

Apparently engineering is not the only profession that neglects telling the public what it can have. Here is an architect with a message for architects *and* engineers.

> If our conception of America the Beautiful is lost, I suggest that there will be four fundamental reasons:
>
> - Lack of public understanding that things can be better than they are.
> - Continuing misdirection of our burgeoning technology and its applications.
> - Success of vested interests in subordinating the community interest to individual advantage.
> - Catastrophic failure of the design professions to demonstrate what the community can expect.
>
> We in the design professions must extend our strongest efforts to stimulate an aware and enlightened public, a sympathetic government, and an inspired business community to demand good design. If we fail, then we may forever lose the chance to rescue a society that now lives amid unsavory, unpleasant, and unnecessary ugliness.*

Engineers probably know as well as anyone what some of the barriers are to the solution of many techno-social problems. The technical know-how to solve problems like housing and public transportation has been around for a while; the obstacles to action are mainly economic and institutional. We could do wonders for the price and supply of housing *if* it weren't for restrictive unions, tradition-bound builders, and outmoded building codes. If the public wants but doesn't get something that engineers can provide, we should lay it on the line, point out what the obstacles are, institutional or otherwise.

We have taken it on the chin because supposedly too few of us are willing to devote our talents to problems of high social significance. Bah; there are plenty of engineers who would like jobs in which they can do something more significant for mankind by contributing to the solution of urban problems, pollution control, and health care problems, for instance. But in spite of all the *talk* about these problems by politicians, the kind of money that's needed to do to them what we have done to our space "problems" isn't there. This is a matter of national priorities and

*Morris Ketuchum, Jr., "AIA's 'War on Community Ugliness,'" in *Consulting Engineer,* January 1966.

of their conversion to effective programs of action. *It is not a case of indifferent engineers.* Our big failing here is our reluctance to state publicly, "Don't tell us, tell your Congressmen."

*"There's always a price to be paid. The engineer rarely makes the point, just as the public often forgets it."** Periodically, people must be reminded that, outside of schoolbooks, there are no such things as perfect solutions to problems. Government serves many useful functions but, in order to be useful, it must restrict the freedom of men; paper is a remarkably useful solution to man's communication problems, but it does take away from the woodlands; houses occupy what once were attractive meadows. In each case there is a price to be paid yet, as a consequence, would you seriously propose doing away with government or paper or houses? It is no different for *any* of man's contrivances. So why are people demanding technical solutions that are perfectly safe or have no detrimental effects on the environment or displace no one? We should speak up to remind our fellow men that perfection is unattainable, and near perfection is prohibitively expensive. It will never be any different.

We fail in another way—we are too modest about our accomplishments. During one week of the year—National Engineers' Week—we make some attempt to call the nation's attention to our achievements, and that's about it. But let's face it; the public is quick to take for granted what *we* take for granted as everlastingly self-evident.

And Now for Your Part

You have been reading about institutional and professional reforms that are needed if something is to be done about biases and ignorances associated with modern technology. Frankly, remedies are not going to evolve spontaneously. Pressure must be brought to bear on government, educational institutions, the media, and the techno-scientific professions. Hopefully, you will add your encouragement.

Hopefully, too, *you* will become and remain informed about society's major technical problems, issues, and opportunities. Hopefully, *you* will take an interest in the decisions your government makes in this area *and* the manner in which it is making them. Hopefully, *you* will think as well as read—bias and prop-

* "Making Tomorrow Happen," published by the Engineers' Public Information Council.

aganda show no signs of becoming extinct. Hopefully, *you* will speak out when speaking out is called for.

EXERCISES

1. Suppose there was a daily "Engineering for the Citizen" column syndicated to newspapers. What *types* of topics, questions, issues, and the like do you feel it could cover? How about a list of titles (specific topics) for 10 of these columns?

2. You be a guest contributor to a syndicated "Engineering for the Citizen" column. Here are some topics, but you may well come up with something better: "Clean Air or Cleaner Air?"; "Is Nuclear Energy Safe? versus How Safe is Nuclear Energy?"; "National Data Bank—Promise, Threats, Issues"; "Greater Leisure—Prospects and Problems"; "Waste Disposal versus Waste Management versus Resource Management."

3. On page 322 there is a brief item appearing in *Time* magazine. Is that an objective, even realistic, piece of journalism?

4. Write an editorial message presumably to be published in an engineering journal, chiding engineers for their reluctance to write for Everyman, and inspiring them to be more communicative in this respect.

Annotated Reading List

CASE HISTORIES OF ENGINEERS IN ACTION

The following are detailed reports on the evolution of engineering creations, mostly from the recent past. Similar reports on projects in the distant past are listed under History of Engineering. By far the best source of case histories is the Stanford Engineering Case Library, Room 500, Stanford University, Stanford, California 94305. They have an extensive collection of excellent, in-depth, well-illustrated cases. Your library may carry a bound set. Other case history sources:

1. **Brennan, Jean F., *The Elegant Solution,* Van Nostrand, New York, 1967. A series of in-depth case histories of projects such as the Nautilus submarine, first transatlantic cable, Boeing 707, and the digital computer. (Obviously author Brennan does not use the term *elegant* in the sense that engineers do.)**
2. **Furnas, C. C. and J. McCarthy, *The Engineer,* LIFE Science Library, TIME, Incorporated, New York, 1966. Splendidly illustrated descriptions of a number of major engineering ventures such as the Verrazano-Narrows Bridge.**
3. **Kingery, Berg, and Schillinger, *Men and Ideas in Engineering,* University of Illinois Press, Urbana, Ill., 1967. A series of cases encompassing a wide variety of engineering creations, written with a personal touch that almost enables you to relive the creative experiences described.**
4. **O'Brien, Robert, *Machines,* LIFE Science Library, TIME, Incorporated, New York, 1964. Liberally illustrated descriptions of the development and workings of a variety of significant machines.**
5. **Page, Robert M., *The Origin of Radar,* a Doubleday Anchor Book, Garden City, N.Y., 1962. A case history in**

depth, tracing the evolution and application of modern radar systems. Fascinating, and only $1.25 in paperback.

6. **Post, Dan R., *Volkswagen–Nine Lives Later,* Horizon House, Arcadia, Calif., 1966. It's fascinating. Treats the history, guts, promotion, and just about everything you might like to know about the VW, and rather colorfully. About 500 illustrations.**
7. **Rapport, S. and Wright, H., *Engineering,* Washington Square Press, 1963. A charming book with a historical flavor, with chapters on Leonardo DaVinci, James Watt, nuclear power, automation, Brooklyn Bridge, and many other well-known projects and persons. It's a 60¢ paperback; how can you go wrong?!**
8. **Whinnery, John R., *The World of Engineering,* McGraw-Hill, New York, 1965. In contrast to the Rapport-Wright book, which spans the past and present, this book emphasizes the present and especially the future. There are chapters on energy, resource engineering, large structures, bio-engineering, space engineering, and numerous other areas where much of the "technical action" will be in the future. A modest-priced paperback.**

DESIGN—PROBLEM SOLVING

There is a host of books in the library with *design* in the title but that say absolutely nothing about design procedure. They are devoted to the description and analysis of *the objects designed* (e.g., gears, circuits, bridges). These are not the books you want if you desire more information on the design process in general. Here are some books you can consult if that is what you seek.

1. **Alger, J. and Hays, C. V., *Creative Synthesis in Design,* Prentice-Hall, Englewood Cliffs, N. J., 1964. A practical little book on creativity, with numerous illustrations and problems. Short paperback.**
2. **Asimow, Morris, *Introduction to Design,* Prentice-Hall, Englewood Cliffs, N.J., 1962. Generally considered to be a classic, although it is at the intermediate level. Short paperback.**
3. **Beakley, G. C. and E. G. Chilton, *Design–Serving the Needs of Man,* Macmillan Publishing Company, New York, 1974. Elementary but extremely well-illustrated.**

4. **Hill, Percy H., *The Science of Engineering Design*, Holt, Rinehart and Winston, New York, 1970. It is introductory, well-packaged, and includes a number of cases.**
5. **Koberg, D. and J. Bagnall, *The Universal Traveler*, William Kaufmann, Inc., Los Altos, Calif., 1974. It's a cute, clever paperback book on problem solving by two architects. It is deceptively titled; the following subtitle tells you more: "A Soft-Systems Guide to Creativity, Problem Solving, and the Process of Reaching Goals." For me it was a very refreshing experience.**
6. **Moore, A. D., *Invention, Discovery, and Creativity*, a Doubleday Anchor Book, Garden City, N. Y., 1969. Written by a popular lecturer on the subject of invention, this is an inspiring little book with fascinating examples. Paperback, at $1.45.**
7. **Wilson, I. G. and M. E. Wilson, *From Idea to Working Model*, Wiley, New York, 1970. You wouldn't know it from the title, but this book treats engineering problem solving, and in a rather unconventional and appealing way at that.**
8. **Woodson, Thomas T., *Introduction to Engineering Design*, New York, McGraw-Hill, 1966. Comprehensive, practical, good all-around characterization of the design process and related activities.**

MODELING

None of these are terribly abstract; they are mainly introductory or intermediate level.

1. **Ackoff, Russel L., *Scientific Method*, Wiley, New York, 1962. The title is somewhat misleading, but the coverage of modeling is excellent. Intermediate level.**
2. **Bross, Irwin, D. F., *Design for Decision*, Macmillan, New York, 1953. One chapter is devoted to modeling, but it's good and it's general. Paperback.**
3. **Dixon, John R., *Design Engineering: Inventiveness, Analysis, and Decision Making*, McGraw-Hill, New York, 1966. Part II is an excellent, intermediate-level treatment of modeling, particularly mathematical modeling.**
4. **Forrester, Jay W., *Urban Dynamics*, MIT Press, Cambridge, Mass., 1969. Fascinating description of a compu-**

ter model of an urban system. Popular, controversial, significant, expensive. Especially if you are interested in urban problems and a promising new approach to them, read it.

5. Forrester, Jay W., *World Dynamics,* Wright-Allen Press, Inc., Cambridge, Mass., 1971. A significant and controversial work describing a computerized world population-growth model. Forrester also offers some very sensible words on the evaluation of models, his in particular.
6. Forrester Jay W., *Principles of Systems,* Wright-Allen Press, Cambridge, Mass., 1968. If you are interested in the "guts" of Forrester's method of modeling social systems, this is the book for you. Paperback, but not cheap.
7. Meadows, D., J. Randers, and W. W. Behrens, *The Limits to Growth,* Universe Books, New York, 1972. A best-selling, paperback version of Forrester's world model mentioned above.
8. Morse, Philip M., *Operations Research for Public Systems,* MIT Press, Cambridge, Mass., 1967. Not a technique book; provides numerous case histories demonstrating the application of modeling skills to urban, health care, criminal justice and other nonengineering problems. Paperback.
9. Tocher, K. D., *The Art of Simulation,* Van Nostrand, New York, 1963. Requires a knowledge of statistics and probability theory, but it is a good little book on simulation technique. Not computer oriented; intermediate level.
10. VerPlanck, D. W., and B. R. Teare, *Engineering Analysis,* Wiley, New York, 1954. Age doesn't detract from this excellent description of engineers' modeling methods. Emphasizes mathematical modeling, so you should know your mathematics.
11. Walker, Marshal, *The Nature of Scientific Thought,* Prentice-Hall, Englewood Cliffs, N.J., 1963. Chapter 1, "The Scientific Method," is truly a classic. If you don't read beyond those 13 pages, you have gotten your money's worth ($2.25 in paperback). This is a highly respected philosophical-historical treatment of modeling and its role in the scientific method. Anyone serious about a career in science should read it.

OPTIMIZATION

Many of the books with *optimization* in the title are mathematical enough to make your hair stand on end. Good introductory treatments of the topic are hard to come by, which explains this short list.

1. **Ackoff, Russel L., *Scientific Method,* Wiley, New York, 1962. You would never guess it from the title, but this book offers a superb treatment of optimization.**
2. **Wilson, Warren E., *Concepts of Engineering Systems Design,* McGraw-Hill, New York, 1965. It's only one chapter, but it's a good one.**

KNOWLEDGE ENGINEERS ACQUIRE

Here is a sampling of books that introduce you to the types of knowledge that engineers must acquire to create what they do. The subjects of feedback and computers are treated separately.

1. **Goldwater, Daniel, *Bridges,* Young Scott Books, New York, 1965. Beautifully illustrated book that explains with the aid of clever sketches some of the principles of bridge design. Read this and wow your friends with your "grasp of the technical" (gambling, of course, that they don't ask you to go on and explain magneto-hydrodynamics and a few more like that!).**
2. **Katz, D. L. et al., *Engineering Concepts and Perspectives,* Wiley, New York, 1968. Gives you a good overview of the kinds of technical knowledge that engineers acquire. Some mathematical background will be helpful. Paperback.**
3. **Mott-Smith, Morton, *Principles of Mechanics Simply Explained,* Dover, New York, 1963. How about that? An inexpensive paperback that explains forces, conservation of momentum, acceleration, etc. without mathematics. As the *Herald Tribune* said: "Dr. Mott-Smith has the ability to illustrate an abstruse principle by means of an absurdly simple demonstration."**
4. **Vallentine, H. R., *Water In the Service of Man,* Penguin Books, Baltimore, Md., 1967. This is a fascinating yet inexpensive paperback that explains viscosity, flow in**

pipes and channels, water power, hydraulic models, hydrology, water resources, and water supply systems through history, all in nonmathematical terms.

5. **VanNess, H. C., *Understanding Thermodynamics*, McGraw-Hill, New York, 1969. It's a classic. A little knowledge of mathematics is helpful. If you have to or want to learn something about thermodynamics, this is the way to do it. Paperback.**

DIGITAL COMPUTERS

The following books are recommended because they emphasize computer *applications* and *implications*. If you are interested in computer "guts" or programming, you should look elsewhere. But you won't have to look far, the market is glutted with such books.

1. **Dorf, Richard C., *Computers and Man*, Boyd and Fraser Publishing Company, San Francisco, 1974. A short introduction to the general workings of computers, followed by an excellent survey of computer applications. Nicely illustrated; paperback.**
2. **Fetter, William A., *Computer Graphics in Communication*, McGraw-Hill, New York, 1964. Although it is an older book, it is an excellent introduction to computer graphics, including computer art. Paperback.**
3. **Gotlieb, C. C. and A. Borodin, *Social Issues in Computing*, Academic Press, New York, 1973. A rather scholarly treatment of computer applications and especially implications. Also addressed to major issues such as invasion of privacy, effects on unemployment, and impact on values.**
4. **Kemeny, John G., *Man and the Computer*, Charles Scribner's Sons, New York, 1972. A highly respected, scholarly treatment of the computer's capabilities, applications, and implications.**
5. **Kochenburger, R. J. and C. J. Turcio, *Computers in Modern Society*, Wiley, New York, 1974. Much of this work is on the nature and workings of computers, and it is beautifully done. About one-fourth is on applications.**
6. **Rothman, S. and C. Mosmann, *Computers and Society*, Science Research Associates, Chicago, 1972. A well-**

rounded, well-illustrated book on workings, applications, and issues.

7. **Sanders, Donald H., *Computers in Society*, McGraw-Hill, New York, 1973. A good introduction to the general workings of computers and to applications in fields such as law, medicine, education, and business.**
8. **Snyder, G. S., *The Computer*, Books by U. S. News and World Report, 1972. For a survey of computer uses, you can't beat it or the price. Paperback, $3.**

FEEDBACK SYSTEMS

If you want to learn more about feedback systems—and it is a fascinating and pervasive topic—you should also look for titles that include *control* or *cybernetics*. (Cybernetics is an in-vogue term usually interpreted to mean control systems *in general*. It derives from a Greek word meaning steersman or helmsman.) Chances are that if you go into the library and pick a book on this general subject, it will leave you in a "cloud of mathematical dust" after the first page or two. The following books emphasize concepts and illustrations, not the mathematics of control system analysis.

1. **Beer, Stafford, *Cybernetics and Management*, Wiley, New York, 1964. An inexpensive little paperback that is more general than the title implies.**
2. **Mayr, Otto, *The Origins of Feedback Control*, MIT Press, Cambridge, Mass., 1970. Truly a fascinating book. Describes interesting feedback devices through history. Numerous diagrams of pressure, speed, temperature, and other types of regulators.**
3. **Metz, L. D. and R. E. Klein, *Man and the Technological Society*, Prentice-Hall, Englewood Cliffs, N.J., 1973. Chapter entitled "Cybernetics and the Concept of Feedback" is an interesting introduction. Paperback.**
4. **Parsegian, V. L., et al., *Introduction to Natural Science; Part I: The Physical Sciences*, Academic Press, New York, 1968. Chapter 6 is an excellent general discussion of feedback systems. The subject of feedback comes up frequently in Part II, also.**
5. **Savas, E. S., "Cybernetics in City Hall," *Science*, May 29, 1970. A fascinating introduction to the relevance of**

feedback concepts to a political system, in this case, the administration of New York City.

6. *Automatic Control,* Simon and Schuster, New York, 1955. A poorly titled collection of relatively general articles that have appeared in *Scientific American* on the subject of feedback and feedback systems. An inexpensive paperback that remains popular.
7. *Man and His Technology,* McGraw-Hill, New York, 1973. Chapters entitled "Feedback" and "Stability" are excellent introductions.

ENGINEERING—GENERAL

Most of these are introductory, directed to engineers in training.

1. Beakley, G. C. and H. W. Leach, *Engineering: An Introduction to a Creative Profession,* 2nd ed., Macmillan, New York, 1972. Written for engineering freshmen, it covers the waterfront; everything from the details of measurement on up.
2. Furnas, C. C. and J. McCarthy, The *Engineer,* LIFE Science Library, TIME Incorporated, New York, 1966. Don't underrate it because of its source. It is an excellent book and, of course, it is not shy on illustrations.
3. Glorioso, R. M. and F. S. Hill, *Introduction to Engineering,* Prentice-Hall, Englewood Cliffs, N.J., 1975. A survey of the types of technical knowledge engineers must acquire (e.g., energy conversion principles, fundamentals of electrical circuits). Emphasizes the kinds of systems engineers create (e.g., communication, transportation, bridge, and computer systems). It's more than a picture book; there are many technical details.
4. Metz, L. D. and R. E. Klein, *Man and the Technological Society,* Prentice-Hall, Englewood Cliffs, N.J., 1973. It's an introduction to engineering that parallels Parts 2 and 3 of this book, and so they could have titled it more appropriately.
5. *Man and His Technology,* McGraw-Hill, New York, 1973. A well-packaged book, not about engineers specifically, but indirectly it tells you much about the ways and creations of engineers in an especially interesting fashion.

HISTORY

Books under the titles of history of engineering and history of technology tend to include much the same kind of information, so a sampling of each is provided, with emphasis on paperbacks.

1. **Burstall, Aubrey F., *A History of Mechanical Engineering,* MIT Press, Cambridge, Mass., 1965. Comprehensive (ancient times through 1960), scholarly, profusely illustrated, paperback.**
2. **DeCamp, L. Sprague, *The Ancient Engineers,* MIT Press, Cambridge, Mass., 1963. A detailed account of those remarkable engineers and their equally remarkable achievements. Paperback.**
3. **Finch, James Kip, *The Story of Engineering,* Doubleday Anchor Books, Garden City, N.Y., 1960. An ambitious work available as an inexpensive paperback. Places more emphasis on personages.**
4. **Kirby, R. S., et al., *Engineering in History,* McGraw-Hill, New York, 1956. Organized by topic (e.g., canals, iron and steel, tunnels, electrification). Fascinating paperback.**
5. **Klemm, Friedrich, *A History of Western Technology,* MIT Press, Cambridge, N.Y., 1964. A popular, comprehensive history of developments from Graeco-Roman to modern times. Paperback, $3.**
6. **Kranzberg, M. and C. W. Pursell, *Technology in Western Civilization,* Vols. I and II, Oxford University Press, New York, 1967. A respected and comprehensive work. Better than usual coverage of technological history in the twentieth century.**

TECHNOLOGY AND SOCIETY—THE INTERACTION

A list of books under this rather broad heading could fill a book and in fact has (*Science, Technology and Public Policy–A Selected and Annotated Bibliography,* prepared by L. K. Caldwell, Department of Government, Indiana University, Vol. I, 1968; Vol. II, 1969; Vol. III, 1972). What follows can only be a small fraction of what is in print.

One source of books, reports, and articles warrants special

mention: the Harvard University Program on Technology and Society. It has produced more publications than it is feasible to list. Although the program itself is defunct, publications can be obtained from the Harvard Information Office, 1350 Massachusetts Avenue, Holyoke Center, Cambridge, Mass. 02138.

General

1. **Allen, F. R., et al., *Technology and Social Change*, Appleton-Century-Crofts, New York, 1957. Don't let the copyright date discourage you; on a subject like this it matters little. Furthermore, this is a comprehensive analysis of the effects of technology on various institutions that makes very useful reading to this day. Written for a college-level course.**
2. **deNevers, Noel, *Technology and Society*, Addison-Wesley, Reading, Mass., 1972. A well-rounded collection of readings focused on the implications of technology and major issues. Prepared especially as a basis for discussion. Paperback.**
3. **Dorf, Richard C., *Technology, Society, and Man*, Boyd and Fraser Publishing Company, San Francisco, 1974. It's excellent, and written by an engineer, too.**
4. **Knelman, Fred H., *1984 and All That*, Wadsworth, Belmont, Calif., 1971. Another book of readings with an impressive list of contributors writing on technology, its implications, and its direction and control. Provocative; available in paperback.**
5. **Muller, Herbert J., *The Children of Frankenstein*, Indiana University Press, Bloomington, Ind., 1970. Reflections on technology's social and cultural implications. It provides historical background plus an examination of technology's effects on war, science, government, business, language, education, and so forth. Considering Professor Muller's reputation as scholar and author, the 400-plus pages, and the paperback price of $3, it is a good buy.**
6. **Teich, Albert H., *Technology and Man's Future*, St. Martin's Press, New York, 1972. Still another book of readings, in this case offering an unusual opportunity to sample the words of technophobes and technophiles. It also dwells on technology assessment. Paperback.**

Impact Studies

1. Cottrell, Fred, *Energy and Society,* McGraw-Hill, New York, 1955. An ever-more relevant treatise on energy —its forms, conversion, utilization, and distribution, and its political, economic, and social implications. Some historical content; scholarly; available in paperback.
2. Levy, Lillian, *Space: Its Impact on Man and Society,* Norton, New York, 1965. A collection of papers evaluating the impact of present and future space exploration. Contributors don't always agree, which doesn't surprise you.
3. Morison, Elting E., *Men, Machines, and Modern Times,* MIT Press, Cambridge, Mass., 1966. A series of historical accounts of the evolution and impact of selected technological innovations like Bessemer steel, continuous-aim firing in the U. S. Navy, and pasteurization of milk.
4. Street, James H., *The New Revolution in the Cotton Economy,* University of North Carolina Press, Chapel Hill, N.C., 1957. An excellent, detailed case study of the development, diffusion, and impact—social and economic—of mechanization.
5. Trachtenburg, Alan, *Brooklyn Bridge: Fact and Symbol,* Oxford University Press, New York, Interesting, scholarly treatment of the impact on the American imagination of one of the great technical feats of the nineteenth century.

Technology and Government

1. Art, Robert J., *The TFX Decision,* Little, Brown and Company, Boston, Mass., 1968. A thorough, well-documented analysis of key decisions concerning a military aircraft. It's a revelation—unless you are intimately familiar with decision making in the Pentagon. Paperback.
2. Carter, Luther J., "The McNamara Legacy: A Revealing Case History—Death of the B-70 Bomber," *Science,* February 23, 1968. It is revealing, especially of the heavy lobbying a federal agency will indulge in on behalf of a major technical project.
3. Lakoff, Sanford A., *Knowledge and Power,* the Free

Press, Collier-Macmillan, New York, 1966. Although this book contains some essays on the technology-government interaction, the main attraction is a series of detailed accounts of major techno-political controversies (e.g., project Mohole, Comsat, establishment of NASA, and the nuclear test-ban treaty).

4. Mandelbaum, Leonard, "Apollo: How the United States Decided to Go to the Moon," *Science*, February 14, 1969. If you can't take horror stories concerning government decision making, don't read it.
5. Nieburg, Harold L., *In the Name of Science*, Quadrangle, New York, 1966. An attack on the federal government and the complex of business and educational institutions that feed on and influence government's major technical ventures. He makes some startling revelations and accusations that you would probably like your congressman to read.
6. *Technical Information for Congress*, 1969, Report to the Subcommittee on Science, Research, and Development of the Committee on Science and Astronautics, U.S. House of Representatives. If you are interested in detailed accounts of major techno-political controversies in Congress, order it from the Superintendent of Documents. Paperback, $2.25.

COMMENTARIES

These were selected primarily because they reflect attitudes toward technology, its role, and its use and misuse. Mainly, this is a sampling of the voice of the technophobe.

1. Butler, Samuel, *Erewhon*, originally published in 1872 but now available in paperback, Airmont Classics Series or Signet Classics, New American Library, New York, 1960. A classic satirical novel, a social commentary on the times, but with an interesting twist that explains its mention here. In a fictional land called Erewhon all machines are banned, out of fear of their ability to become dominant over man.
2. Ellul, Jacques, *The Technological Society*, Alfred Knopf, New York, 1964. A controversial work widely viewed as a pessimistic view of man's future in a technological world.

It is herein that he cites his often quoted characterization of technology as an autonomous, self-generative force.

3. Ferkiss, Victor C., *Technological Man,* a Mentor Book from New American Library, New York, 1969. Among other virtues, this book summarizes and analyzes the views of a number of technophobic philosophers in a refreshingly objective fashion. It also offers some special insights into the present and probable future implications of technology.
4. Huxley, Aldous L., *Brave New World,* published in 1932 by Harper and Row, New York, now available as a Bantam Modern Classic paperback, Bantam Books, New York, 1953. A satirical novel describing a mechanized utopia in which Ford and his standardized Model T are worshipped. As a social commentary, it has received considerable attention to this day, but never drew raves as a novel.
5. Maddox, John, *The Doomsday Syndrome,* McGraw-Hill, New York, 1972. A timely, well-informed response to a flood of literature emanating from the doomsday prophets, particularly of the environmental and population-growth variety.
6. Marine, Gene, *America the Raped,* Simon and Schuster, New York, 1969. A caustic indictment of public and private decision makers who, according to Marine, are destroying the environment. It is a biting attack on engineers, as the following subtitle of his book indicates: "Engineering Mentality and Devastation of A Continent." It is also an unusual exhibition of political and organizational naiveté.
7. Mumford, Lewis, *The Myth of the Machine, Vol. 2; The Pentagon of Power,* Harcourt, New York, 1970. A rather eloquent and vigorous attack on technology and especially misuses of it. Known for his technophobic bias and pessimism, this, Mumford's latest book, is no exception to the pattern he has established.
8. Roszak, Theodore, *The Making of a Counter Culture: Reflections on the Technocratic Society and Its Youthful Opposition,* Doubleday, New York, 1969. An elaborate critique of our "technocratic society," plus a sampling of views of Marcuse, Goodman, and others of like mind, plus an interesting account of the rise of a counterculture. Available in paperback.

KEEPING INFORMED

These are periodicals known for their coverage of major technological issues of the times. These are not the only journals that perform a service in this respect. These are listed because you are less likely to be aware of their reputations for superior coverage of technical issues.

1. ***Science,*** **official publication of the American Association for the Advancement of Science. True, it covers a broad spectrum, but it is a weekly, and** ***some*** **portion of every issue is devoted to technical issues. The quality of techno-scientific journalism has earned worldwide respect.**
2. ***Technology Review,*** **published at MIT, gives excellent coverage to major technical problems, issues, and developments. It is not the esoteric publication you might expect, considering its source; it is quite readable and enjoys an excellent reputation for its high-caliber techno-scientific journalism.**
3. ***Technology and Culture,*** **journal of the Society for the History of Technology. It is a scholarly journal devoted to matters contemporary as well as past.**
4. ***Impact of Science and Society,*** **a UNESCO publication, available from the UNESCO Publications Center, P. O. Box 433, New York, N.Y. 10016. Focuses on the impact of science and technology, spanning a broad range of topics.**
5. ***The Futurist,*** **official publication of the World Future Society, an organization of persons interested in study of the future. Inevitably, many of the articles in such a publication will concern technology's future directions and implications. It offers a fascinating reading experience.**
6. ***The Bulletin of the Atomic Scientists,*** **a first-rate journal on science, engineering, and public affairs, which is not restricted to the subject of atomic energy.**

PICTURE CREDITS

Western Electric Company
1-3, 1-4

General Electric Company
1-5

Chesapeake Bay Bridge-Tunnel District
1-6 to 1-11

Teledyne Ryan Aeronautical Company
1-12 to 1-14

Lockheed-California Company
1-15

Teledyne Ryan Aeronautical Company
1-16

British Hovercraft Corporation
2-1

National Aeronautics & Space Administration
2-2

The Cleveland Bridge and Engineering Company Limited, Great Britain
3-1*a*

M. W. Martin, M.W.M. Enterprises, Inc.
3-1*b*

British Petroleum Company, Ltd.
3-2*a*

Anderson and Anderson, Consulting Engineers, and the National Science Foundation
3-2*b*

United States Navy
3-2*c*

Craven, Kikutake, St. Denis, Pryor, Wilkins, & Burgess
3-2*d*

Scripps Institution of Oceanography
3-2*e*

National Aeronautics and Space Administration
3-3

Aerovironment, Inc. The device, trade-named AeroBoost, invented by P. B. MacCready and P. B. S. Lissaman
3-5

Cutler-Hammer Inc.
3-6*a*

Lockwood Corporation, Gering, Nebraska
3-6*b* and 3-6*c*

Environmental Research Laboratory, University of Arizona
3-7

Aerojet-General Corporation
3-9

LTV Aerospace Corporation
5-1*a* and 5-1*b*

Hawker Siddeley Aviation Limited
5-1*c*

Dornier GMBH
5-1*d*

Avions Marcel-Dassault-Bereguet Aviation
5-1*e*

Hughes Helicopter Division of Summa Corporation
5-1*f*

Sikorsky Aircraft Division of United Aircraft Corporation
5-1*g*

Textron's Bell Aerospace Company
5-1*h*, 5-1*i*, and 5-1*j*

Society Nationale Industrielle Aerospatiale, France
5-1*k*

San Francisco Bay Area Rapid Transit District
6-4

Photo-archives of the Helsinki City Office, copyright by Lehtikuva Oy
6-5

The Warner and Swasey Company
6-6

Air Products and Chemicals Inc.
9-1 and 9-3

Dr. J. E. Cermak, Fluid Dynamics and Diffusion Laboratory, Colorado State University
9-4

United States Navy
9-5

United States Army
9-6

The Boeing Aerospace Company
9-7

Federal Aviation Agency
9-12

The Boeing Aerospace Company
9-13*a*

EXXON Corporation
9-13*b*

University of California Institute of Transportation and Traffic Engineering, and the Federal Aviation Agency
9-14

Wide World
10-1

Julian Wasser, Time-Life Picture Agency, c Time Inc.
10-2

BASF Systems, Inc.
10-7

IBM Corporation
13-2

Lockheed Missles and Space Company
13-5*a*

General Motors Research Laboratories
13-5*b* and 13-6

General Electric Company
13-7 and 13-8

Burroughs Corporation
13-9

National Astronomy and Ionosphere Observatory, of the National Astronomy and Ionosphere Center, operated by Cornell University for the National Science Foundation
14-2

United States Coast Guard
19-1

Index